Advances in Cognitive Engineering and Neuroergonomics

Advances in Human Factors and Ergonomics Series

Series Editors

Gavriel Salvendy
Professor Emeritus
School of Industrial Engineering
Purdue University

Chair Professor & Head
Dept. of Industrial Engineering
Tsinghua Univ., P.R. China

Waldemar Karwowski
Professor & Chair
Industrial Engineering and
Management Systems
University of Central Florida
Orlando, Florida, U.S.A.

3rd International Conference on Applied Human Factors and Ergonomics (AHFE) 2010

Advances in Applied Digital Human Modeling
Vincent G. Duffy

Advances in Cognitive Ergonomics
David Kaber and Guy Boy

Advances in Cross-Cultural Decision Making
Dylan D. Schmorrow and Denise M. Nicholson

Advances in Ergonomics Modeling and Usability Evaluation
Halimahtun Khalid, Alan Hedge, and Tareq Z. Ahram

Advances in Human Factors and Ergonomics in Healthcare
Vincent G. Duffy

Advances in Human Factors, Ergonomics, and Safety in Manufacturing and Service Industries
Waldemar Karwowski and Gavriel Salvendy

Advances in Occupational, Social, and Organizational Ergonomics
Peter Vink and Jussi Kantola

Advances in Understanding Human Performance: Neuroergonomics, Human Factors Design, and Special Populations
Tadeusz Marek, Waldemar Karwowski, and Valerie Rice

4th International Conference on Applied Human Factors and Ergonomics (AHFE) 2012

Advances in Affective and Pleasurable Design
Yong Gu Ji

Advances in Applied Human Modeling and Simulation
Vincent G. Duffy

Advances in Cognitive Engineering and Neuroergonomics
Kay M. Stanney and Kelly S. Hale

Advances in Design for Cross-Cultural Activities Part I
Dylan D. Schmorrow and Denise M. Nicholson

Advances in Design for Cross-Cultural Activities Part II
Denise M. Nicholson and Dylan D. Schmorrow

Advances in Ergonomics in Manufacturing
Stefan Trzcielinski and Waldemar Karwowski

Advances in Human Aspects of Aviation
Steven J. Landry

Advances in Human Aspects of Healthcare
Vincent G. Duffy

Advances in Human Aspects of Road and Rail Transportation
Neville A. Stanton

Advances in Human Factors and Ergonomics, 2012-14 Volume Set: Proceedings of the 4th AHFE Conference 21-25 July 2012
Gavriel Salvendy and Waldemar Karwowski

Advances in the Human Side of Service Engineering
James C. Spohrer and Louis E. Freund

Advances in Physical Ergonomics and Safety
Tareq Z. Ahram and Waldemar Karwowski

Advances in Social and Organizational Factors
Peter Vink

Advances in Usability Evaluation Part I
Marcelo M. Soares and Francisco Rebelo

Advances in Usability Evaluation Part II
Francisco Rebelo and Marcelo M. Soares

Advances in Cognitive Engineering and Neuroergonomics

Edited by
Kay M. Stanney
and
Kelly S. Hale

CRC Press
Taylor & Francis Group
Boca Raton London New York

CRC Press is an imprint of the
Taylor & Francis Group, an **informa** business

CRC Press
Taylor & Francis Group
6000 Broken Sound Parkway NW, Suite 300
Boca Raton, FL 33487-2742

© 2012 by Taylor & Francis Group, LLC
CRC Press is an imprint of Taylor & Francis Group, an Informa business

No claim to original U.S. Government works

Printed in the United States of America on acid-free paper
Version Date: 20120529

International Standard Book Number: 978-1-4398-7016-7 (Hardback)

This book contains information obtained from authentic and highly regarded sources. Reasonable efforts have been made to publish reliable data and information, but the author and publisher cannot assume responsibility for the validity of all materials or the consequences of their use. The authors and publishers have attempted to trace the copyright holders of all material reproduced in this publication and apologize to copyright holders if permission to publish in this form has not been obtained. If any copyright material has not been acknowledged please write and let us know so we may rectify in any future reprint.

Except as permitted under U.S. Copyright Law, no part of this book may be reprinted, reproduced, transmitted, or utilized in any form by any electronic, mechanical, or other means, now known or hereafter invented, including photocopying, microfilming, and recording, or in any information storage or retrieval system, without written permission from the publishers.

For permission to photocopy or use material electronically from this work, please access www.copyright.com (http://www.copyright.com/) or contact the Copyright Clearance Center, Inc. (CCC), 222 Rosewood Drive, Danvers, MA 01923, 978-750-8400. CCC is a not-for-profit organization that provides licenses and registration for a variety of users. For organizations that have been granted a photocopy license by the CCC, a separate system of payment has been arranged.

Trademark Notice: Product or corporate names may be trademarks or registered trademarks, and are used only for identification and explanation without intent to infringe.

Visit the Taylor & Francis Web site at
http://www.taylorandfrancis.com

and the CRC Press Web site at
http://www.crcpress.com

Table of Contents

Section I: Cognitive Engineering: Fundamentals

1	Human factors in the design of a search tool for a database of recorded human behavior R. Abbott, C. Forsythe, M. Glickman and D. Trumbo, USA	3
2	Augmented navigation: A cognitive approach E. Onal, A. Savoy and S. McDonald, USA	13
3	Change blindness when viewing web pages D. Steffner and B. Schenkman, Sweden	23
4	Qualitative facets of the problem statement H. von Brevern, Switzerland and W. Karwowski, USA	33
5	Overloading disks onto a mind: Quantity effects in the attribution of mental states to technological systems O. Parlangeli, S. Guidi and R. Farina, Italy	43
6	Effectiveness of two-channel bone conduction audio communication system in speech recognition tasks M. McBride, R. Weatherless and T. Letowski, USA	53
7	Cognitive ergonomics: The user and brand in an informational relationship S. Oliveira, A. Heemann and M. Okimoto, Brazil	61

Section II: Cognitive Engineering: Workload and Stress

8	Cognitive workload analysis and development of a stress index for life science processes M. Swangnetr, Thailand, B. Zhu and D. Kaber, USA K. Thurow, N. Stoll and R. Stoll, Germany	71
9	Interaction effects of physical and mental tasks on auditory attentional resources A. Basahel, M. Young and M. Ajovalasit, UK	81
10	Applicability of situation awareness and workload metrics for use in assessing nuclear power plant designs S. Guznov, L. Reinerman-Jones and J. Marble, USA	91

11	Effect of stress on short-term memory among engineering workshop trainees B. Keshvari and M. Abdulrani, Malaysia	99

Section III: Cognitive Engineering: Activity Theory

12	On the relationship between external and internal components of activity W. Karwowski, USA, F. Voskoboynikov, Russia, and G. Z. Bedny, USA	109
13	Positioning actions' regulation strategies G. Bedny and W. Karwowski, USA	116
14	Emotional-motivational aspects of a browsing task H. von Brevern, Switzerland, and W. Karwowski, USA	125
15	Using systemic approach to identify performance enhancing strategies of rock drilling activity in deep mines M.-A. Sanda, Sweden/Ghana, J. Johansson, B. Johansson and L. Abrahamsson, Sweden	135

Section IV: Cognitive Engineering: Error and Risk

16	Challenges in human reliability analysis (HRA): A Reflection on the Accident Sequence Evaluation Program (ASEP) HRA procedure H. Liao, A. Bone, K. Coyne and J. Forester, USA	147
17	Mirror neuron based alerts for control flight into terrain avoidance M. Causse, J. Phan, T. Ségonzac and F. Dehais, France	157
18	Computer technology at the workplace and errors analysis I. Bedny, W. Karwowski and G. Bedny, USA	167
19	A study on document style of medical safety incident report for latent factors extraction and risk-assessment-based decision making K. Takeyama, T. Akasaka, T. Fukuda, and Y. Okada, Japan	177

Section V: Cognitive Engineering: Applications

20	Promoting temporal awareness for dynamic decision making in command and control S. Tremblay, F. Vachon, R. Rousseau and R. Breton, Canada	189

21	Data-driven analysis and modeling of Emergency Medical Service process L. Wang and B. Caldwell, USA	199
22	Age and comprehension of written medical information on drug labels M. Chafac, USA, and A. Chan, Hong Kong	207
23	Designing virtual reality systems for procedural task training N. Gavish, Israel	218

Section VI: Neuroergonomics: Measurement

24	Given the state, What's the metric? M. Carroll, K. Stanney, C. Kokini and K. Hale, USA	229
25	Development of a neuroergonomic application to evaluate arousal D. Gartenberg, R. McGarry, D. Pfannenstiel, D. Cisler, T. Shaw and R. Parasuraman, USA	239
26	fNIRS and EEG study in mental stress arising from time pressure S.-Y. Cheng, C.-C. Lo and J.-J. Chen, Taiwan	249
27	Availability and future prospects of functional near-infrared spectroscopy (fNIRS) in usability evaluation H. Iwasaki and H. Hagiwara, Japan	259
28	Team neurodynamics: A novel platform for quantifying performance of teams G. Raphael, R. Stevens, T. Galloway, C. Berka and V. Tan, USA	269

Section VII: Neuroergonomics: Applications

29	Challenges and opportunities for inserting neuroscience technologies into training systems R. Stripling, USA	281
30	A neuroergonomic evaluation of cognitive workload transitions in a supervisory control task using Transcranial Doppler Sonography K. Satterfield, T. Shaw, R. Ramirez and E. Kemp, USA	290
31	From biomedical research to end user BCI applications through improved usability of BCI++ v3.0 P. Perego, E. Gruppioni, F. Motta, G. Verni and G. Andreoni, Italy	300

32 An ERP study on instrument form cognition.doc 308
   Y. Jiang, J. Hong, X. Li and W. Wang, China

Index of Authors 319

Preface

This book brings together a wide-ranging set of contributed articles that address emerging practices and future trends in cognitive engineering and neuroergonomics. Both aim to harmoniously integrate human operator and computational system, the former through a tighter cognitive fit and the latter a more effective neural fit with the system. The chapters in this book uncover novel discoveries and communicate new understanding and the most recent advances in the areas of workload and stress, activity theory, human error and risk, and neuroergonomic measures, as well as associated applications.

The book is organized into two main sections, one focused on cognitive engineering, with 5 subsections, and the other on neuroergonomics, with 2 subsections, including:

Cognition Engineering

    I. Cognitive Engineering: Fundamentals
    II. Cognitive Engineering: Workload and Stress
    III. Cognitive Engineering: Activity Theory
    IV. Cognitive Engineering: Error and Risk
    V. Cognitive Engineering: Applications

Neuroergonomics

    VI. Neuroergonomics: Measurement
    VII. Neuroergonomics: Applications

Sections I through V of this book focus on cognitive engineering, with emphasis on the assessment and modeling of cognitive workload and stress, practical applications of Systemic-Structural Activity Theory, and error and risk management. These chapters challenge the reader to extend current concepts by: considering such constructs as the vector 'motive(s) → goal' within human-system couplings, enhancing guidance for accident avoidance – such as through new types of neurally-based visual alerts and incident report support systems, and developing new interaction techniques such as bone conduction listening devices.

Sections VI and VII uncover some of the latest measurement and application trends in neuroergonomics. The measurement section uncovers linkages between physiological measures and affective and cognitive states, including real-time measures of arousal, hemodynamic changes in the prefrontal cortex that correspond to task difficulty, combinations of fNIR and EEG that can be used to assess the mental stress associated with time pressure, and real-time measures of team cognitive state. The applications section examines the challenges and opportunities for inserting neuroscience technologies into training systems, evaluates the benefits of neuroergonomics for supervisory control tasks, introduces the latest in brain-computer interface systems, and explores the application of physiological data to the design of instrument displays.

Collectively, the chapters in this book have an overall goal of developing a deeper understanding of the couplings between external behavioral and internal mental actions, which can be used to design harmonious work and play environments that seamlessly integrate human, technical, and social systems.

Each chapter of this book was either reviewed or contributed by members of the Cognitive & Neuroergonomics Board. For this, our sincere thanks and appreciation goes to the Board members listed below:

H. Adeli, USA	T. Marek, Poland
G. Bedny, USA	J. Murray, USA
A. Burov, Ukraine	A. Ozok, USA
P. Choe, China	O. Parlangeli, Italy
M. Fafrowicz, Poland	R. Proctor, USA
X. Fang, USA	P. Rau, China
Q. Gao, China	A. Santamaria, USA
P. Hancock, USA	A. Savoy, USA
K. Itoh, Japan	D. Schmorrow, USA
D. Kaber, USA	N. Stanton, UK
W. Karwowski, USA	K. Vu, USA
H. Liao, USA	T. Waldmann, Ireland
S.-Y. Lee, Korea	
Y. Liu, USA	L. Zeng, USA

It is our hope that professionals, researchers, and students alike find the book to be an informative and valuable resource; one that helps them to better understand important concepts, theories, and applications in the areas of cognitive engineering and neuroergonomics. Beyond basic understanding, the contributions are meant to inspire critical insights and thought-provoking lines of follow on research that further establish the fledgling field of neuroergonomics and sharpen the more seasoned practice of cognitive engineering. While we don't know where the confluence of these two fields will lead, they are certain to transform the very nature of human-systems interaction, resulting in yet to be envisioned designs that improve form, function, efficiency, and the overall user experience for all.

April 2012

Kay M. Stanney and Kelly S. Hale
Design Interactive, Inc.
Orlando, FL USA

Editors

Section I

Cognitive Engineering: Fundamentals

CHAPTER 1

Human Factors in the Design of a Search Tool for a Database of Recorded Human Behavior

Robert G. Abbott, Chris Forsythe, Matthew Glickman, Derek Trumbo

Sandia National Laboratories
Albuquerque, NM, USA
Correspondence: rgabbot@sandia.gov

ABSTRACT

Many enterprises are becoming increasingly data-driven. For example, empirically collected data about customer behavior offers an alternative to more traditional, synthetic techniques such as surveys, focus groups, and subject-matter experts. In contrast, recordings of tactical training exercises for the US military are not broadly archived or available for analysis. There may be great opportunity for military training and planning to use analogous techniques, where tactical scenarios are systematically recorded, indexed, and archived. Such a system would provide information for all levels of analysis, including establishing benchmarks for individual performance, evaluating the relevance and impact of training protocols, and assessing the utility of proposed systems and conops. However, such a system also offers many challenges and risks, such as cost, security, privacy, and end-user accessibility. This paper examines the possible benefits and risks of such a system with some emphasis on our recent research to address end-user accessibility.

INTRODUCTION

Human behavior is recorded and analyzed for many different applications. The digitization of record-keeping lead almost immediately to studies of consumer behavior which now flourishes with large-scale analysis of web browsing and online shopping by Google, Facebook, and others. Recording discrete transactions (such as withdrawing money from a bank account or visiting a web page) is relatively straightforward. It is more difficult to track and analyze behaviors with continuous aspects, such as motion in sports or combat tactics. Nonetheless, great advances in technology and infrastructure for tracking have occurred in the relatively recent past. These include the Global Positioning System (GPS), pervasive cellphone networks, and cheap, powerful computer vision implementations such as Microsoft Kinect (Stone 2011). Applications for these technologies include analysis of football tactics (Xu 2005), tracking the foraging of wild elk in Oregon (Kie 2004), and detecting "anomalous" behavior in parking garages and airports (Micheloni 2005).

Our research focuses on weapons and tactics training for defense applications. Training missions involve numerous aircraft, ships, boats, etc (Figure 1). These may be *live* (actual aircraft and ships instrumented with GPS and data downlinks), *virtual* (individually controlled by students in high-fidelity cockpit simulators), or *constructive* (a single person controls numerous lower-fidelity entities using a semi-automated forces tool). *LVC* exercises blend Live, Virtual, and Constructive entities. Significant examples include Fleet Synthetic Training (Jay 2008) and Virtual Flag (Noel 2009). In certain cases, exercises have been conducted primarily as studies, rather than training exercises, but such large-scale exercises are too expensive to conduct on a routine basis. For example, the Millennium Challenge cost $250M (Naylor 2002).

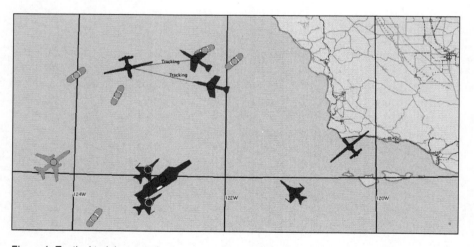

Figure 1: Tactical training exercises connect numerous ships and aircraft, both real and simulated. Reviewing a recording of the scenario during after-action review helps students learn. A large collection of recorded scenarios could be a rich source of knowledge with many applications, but there are challenges in collecting the data and making it accessible to end-users.

Typically individual scenarios are recorded to provide feedback to the participants immediately afterwards. Then the scenario is deleted. Retaining the

data and providing uniform, user-friendly access to it would enable many questions to be studied empirically, such as:
- Understanding individual predictability / variability of students
- Training fidelity studies – compare live vs. virtual vs. constructive tactics
- Combat planning factors – estimate resources to do a job
- Tradespace studies for system designs – determine what tools/systems are needed to reliably achieve a given objective.
- Autonomous systems based on tactics demonstrated by humans

Given a large repository of recorded scenarios, it's necessary to have some practical way to find behavior examples relevant to the task at hand. Algorithms for identifying behaviors from behavior trace data are well-studied, and are surveyed in (Hu 2004), but in most cases these research systems emphasize raw capability without much regard to supporting non-technical end users. Our research focuses on the end-user human/computer interactions such as formulating queries for behavior recognition systems. In previous studies we have measured the accuracy of example-based queries for finding events of interests in these datasets (Stevens 2009), quantified training benefits for students who reviewed key events in training scenarios (Stevens 2010), and created software agents by modeling recorded examples (Abbott 2010). This paper describes our progress in expanding the scope of these techniques from analyzing a single scenario recording towards accessing a potentially large repository of recorded scenarios.

However, before reviewing our own technical progress in operationalizing such repositories, we discuss a set of broader concerns and issues surrounding the collection of such data.

CONSOLIDATING DATA SERVICES

The office of the US Chief Information Officer has issued a Federal Cloud Computing Strategy (Kundra 2011). The goal of the strategy is to increase efficiency by consolidating expertise in providing each service, reduce duplication of effort at individual sites, and increase utilization of computing resources through pooling. This is relevant to archiving scenario recordings because most sites do not have the storage space or dedicated personnel to collect and manage an archive.

On the other hand, a single "global" archive cannot meet the needs of all users. If nothing else, the data must be partitioned into subsets for varying levels of security classification. In some cases security issues might prevent a site from using the service at all unless it can be hosted and managed locally.

Because exercises are already networked, the protocols used for communication (HLA, DIS) provide a technical foundation and also organizational principles for standardization of scenario recordings. Various levels of standards and integration are possible.
- Point Solutions – each site implements its own unique policies using a unique set of tools. Integration is post-hoc (and unlikely to occur).

- Standard Protocols – many different implementations exist but conform to a specification for data exchange. Examples: TCP/IP, CORBA, email.
- Separate Instances of a Single System – sites can exchange data because they use the same software.
- Centralized Service – all sites access a pool of computing resources. Changes to the centralized service are visible to all sites automatically.

It is possible to start small whether working from the top down or bottom-up; a centralized repository might be established and then gradually grow as more users adopt it, or decentralized users might begin with their own repositories and pool them over time. Point solutions are the only situation from which it is very difficult to recover.

Commercial Software Support

Conventional database applications record factual information such as names, addresses, student grades, orders placed. etc. Collecting finer-grained numeric information about specific behaviors (such as tactics) requires making database queries concerning time and space. The Open Geospatial Consortium (http://www.opengeospatial.org) has developed a set of 30 standards for using geospatial information for storing, specifying, and exchanging geospatial information. All major database vendors support geospatial extensions, including Teradata, IBM DB2 and Informix, and Oracle. In total, over 400 private and government organizations belong to the consortium. Software licensing costs are not a barrier to supporting geospatial functionality because the functionality is available in open-source products, namely PostGIS (http://postgis.refractions.net) geospatial extensions for the PostgreSQL database.

PRIVACY AND SECURITY

The objective of collecting data is to improve the performance of individuals and the overall system. However, there is potential harm in exploiting information, both to individuals and the enterprise.

Privacy

Appel recounts the history of a program to assess bombing accuracy in combat missions during the Vietnam War (Appel 1975):

> It was obvious in this initial study that bomb impacts could be plotted with a great deal of accuracy... The evaluation was direct and straightforward and led to a preliminary estimate of a combat CEP. Surprisingly, this preliminary estimate, which contained only 31 measurements, differed only eight percent from a more conclusive CEP derived some 500 samples later...

PACAF took over management of the evaluation... the governing directive, PACAFM 55-25, required that reports be identified by crew member and that wings keep a record of individual aircraft commanders' CEP's. The thrust of the program was clearly changed from one of analysis of accuracy to that of command and control. The reaction from the wings was almost instantaneous. The program came to a virtual standstill, partly because of its new complexity but primarily because the finger was now being pointed at the pilot. Where reports had been anonymous in the past, records were now to be kept on individual accuracy. Cooperation, essential for the plotting of DMPI's, died, and the data submitted dropped to very low levels, both unreliable and unacceptable.

As a result of the impending demise of the program and the importance attached to its original goal of determining accuracy, steps toward simplification were taken, and the reports once again were made anonymous. More was found out about bombing accuracy than was ever anticipated by those who conceived the idea of trying to measure it. Combat planning factors were revised to reflect the newly documented information. Beyond that, though, serious questions were raised regarding our avionics, weapons, and training... Improvements were made, and they were effective.

The main lesson here is that people will rebel against data collection if it could be used against them. The issue is partly cultural; the negative reaction might be less among people employed from the beginning without the expectation of privacy in their job performance. Still, the technology must be designed to implement policy decisions such as reasonable privacy protections, for example by anonymizing the data so names are not associated with behavior traces.

However, making data truly anonymous (so individuals cannot be re-identified even by a determined effort) is far more difficult than might be assumed. Simply removing names, ID numbers, etc. is not effective because the combination of other, seemingly benign pieces of information is highly likely to be unique when taken together, and combined with other datasets. Research on anonymization is discouraging: "Our results demonstrate that even modest privacy gains require almost complete destruction of the data-mining utility" (Brickell 2008). Other researchers dispute that negative conclusion, but the operational community for a given platform is relatively small and close-knit, making anonymity guarantees all but impossible. It will fall to those in authority (i.e. the command structure) as a matter of policy to be clear when data will be used for assessments, vs. when it is "safe" for students to learn by exploring and possibly making mistakes.

Security

The collection of data also poses risks to organizations. Of course, any data collection and archival must conform to existing rules, for example by transmitting data only through approved channels. Even so, systematic collection and archival of data creates a potential windfall for adversaries. After 9/11 the US government was criticized for failing to assemble the information it collected which, taken together, paints a seemingly clear picture of the impending attack (in retrospect). As a result internal firewalls were removed and agencies shared information more freely. Years later, when a large number of documents were released through Wikileaks, critics were quick to criticize the availability of so much information to analysts in seemingly unrelated areas.

Because the pendulum of compartmentalization swings back and forth, a repository must be designed to support whatever policy is in place at the time. Enforcing technical compatibility between systems that are not currently connected would require organizational discipline but would allow the use of common software products and technical experts.

FINDING SITUATIONS OF INTEREST

Formulating queries for tactical behavior is challenging because situations are complex, and the search may be narrowed by several different types of constraints, which suggest different query mechanisms. Table 1 lists types of patterns in a database of tactical behavior, generally in order of increasing sophistication and complexity. This list of capabilities guides our research and may form a set of fea-tures or requirements for tools accessing a common repository of scenario recordings.

Table 1: Types of patterns identified in a database of tactical behavior

Type of Pattern	Examples	Query Specifications
Spatial	Formation	Sketch
		Similarity to previous example
Spatio-Temporal	Flight path	Path sketch
	Tactical movement around an obstacle	Similarity to previous example with start and end times
Object Attributes	Vehicle type	Select from list
	Domain (air/land/sea)	Commonality among multiple previous examples
Action or Change-of-State	Shot fired	Select from list
	Touching the runway	Detect differences in examples from before and after
Sequence	Takeoff, perform check-in, then join formation	Narrative
		Timeline
		Branching timeline (state machine)
Intent	Hold territory	Statistics on sub-patterns over time
	Suppress Air Defenses	Manual tagging

Recent AEMASE Developments - Relational Modeling

For handling a very small number of entities, for example a 2v2 air engagement, the number of entities is known ahead of time and each one plays a distinct role, so there is no question of what data to analyze. This was the focus of our early work on Automated Expert Modeling for Automated Student Analysis (AEMASE) (Abbott 2006). In contrast, a large scenario includes a large number of entities which varies over time. If they are interchangeable for certain purposes, it is necessary to reason about them using categories (or classes). For example, we might be interested in all cases in which any of our aircraft came near any air defense system. Any of a number of objects can fill each of the roles in a given situation.

To meet this requirement, AEMASE is now built on a relational database that represents a scenario as discrete events and timeseries data. Each timeseries or discrete event is described by a set of one or more tags. A tag is a *name=value* pair. For example the position of an aircraft over time might have the tags `domain=air, side=blue, type=f18, callsign=bat-11`. Most situations of interest involve two or more entities and some constraint on their relative positions or headings.

The information that constitutes a Situation (i.e. a template for a given behavioral pattern of interest) is summarized in the Details panel (Figure 2). This panel shows the situation name, the types of entities involved, which properties characterize the situation (e.g. range, altitude, etc.), and examples (and counter-examples) of the situation that have been indicated by the user. Each piece of information can be edited to refine the definition of the situation. Together, these pieces of information define a pattern for similar situations that might occur at other times and involving other entities in the scenario. AEMASE uses machine learning algorithms (e.g. support vector machine) to characterize the spatial relations among entities in the positive vs. negative examples. It then queries a scenario to find all sets of entities that match the tags, and applies the spatial relations model to determine whether the set of entities constitutes a situation of interest at the time.

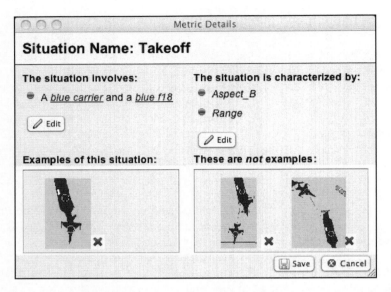

Figure 2: The AEMASE Details dialog summarizes a situation of interest: the name of the situation, types of entities involved, which properties characterize the situation, and examples (and counter-examples) of the situation. In this example the user is looking for planes taking off from blue (friendly) carriers.

Scaling to Collections of Scenarios

The functionality described in the previous section is useful for searching scenarios one at a time for matching situations, but is not sufficient for searching a large collection of scenarios. We are currently defining requirements and developing a concept of operations for searching across multiple scenarios. This will include:
- More powerful searches, implementing more of the hierarchy of patterns in Table 1
- Display only the relevant entities or aspects of each matching situation in each scenario – not a snapshot of the entire scenario
- Handle inconsistency of labeling between scenarios – e.g. "country=U.S." in one scenario, and "nation=USA" in another
- User-specific search patterns – ability to annotate a scenario without changing it in any way visible to other users
- Ability to share patterns and search results when desired
- Retrieval of detailed scenario recording for offline analysis

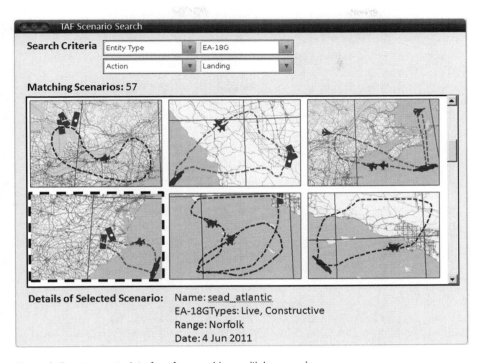

Figure 3: Prototype user interface for searching multiple scenarios

CONCLUSION

It is premature to determine whether enterprise-wide scenarios recordings databases would be cost-justified. End-user demand is not likely to materialize until benefits are demonstrated by compelling proof-of-concept prototypes are developed for a variety of applications. However, incremental improvements in scenario search and analysis will provide additional value with each step, making it an avenue that is worth exploring further.

REFERENCES

R.G. Abbott. "Automated Expert Modeling For Automated Student Evaluation." Intelligent Tutoring Systems 2006: pp. 1-10.

R. G. Abbott, J. D. Basilico, M. R. Glickman, J. H. Whetzel. Trainable Automated Forces. IITSEC 2010. Orlando, FL, 2010.

B. Appel. "Bombing Accuracy in a Combat Environment." Air University Review, July 1975.

J. Brickell and V. Shmatikov. "The cost of privacy: destruction of data-mining utility in anonymized data publishing." In Proc. 14th ACM SIGKDD KDDM, Las Vegas, Nevada, USA, August 24-27, 2008. ACM, 2008., pages 70–78.

N. Getlin. "Virtual Flag 09-3 prepares warfighters for deployment," http://www.nwfdailynews.com/articles/exercise-19408-air-virtual.html August 2009. Accessed 5 Mar 2012.

W. Hu, T. Tan, L. Wang, S. Maybank. "A survey on visual surveillance of object motion and behaviors," Systems, Man, and Cybernetics, Part C: Applications and Reviews, IEEE Transactions on , vol.34, no.3, pp.334-352, Aug. 2004

E.F. Jay. "Fleet Synthetic Training Maximizes At-Sea Training Time." MT2 2008 Volume: 13 Issue: 6 (November/December).

Kie, J.G.; Ager, A. A.; Cimon, N. J.; Wisdom, M.J.; Rowland, M.M.; Coe, P. K.; Findholt, S.L.; Johnson, B.K.; Vavra, M. 2004. The Starkey databases: spatial-environmental relations of North American elk, mule deer, and cattle. In: Transactions of the 69th North American Wildlife and Natural Resources Conference: 475-490.

V. Kundra. "Federal Cloud Computing Strategy," Feb 2011. http://www.cio.gov/documents/federal-cloud-computing-strategy.pdf Accessed 5 Mar 2012.

Micheloni, C.; Salvador, E.; Bigaran, F.; Foresti, G.L.; "An integrated surveillance system for outdoor security," Advanced Video and Signal Based Surveillance, 2005. AVSS 2005. IEEE Conference on , vol., no., pp. 480- 485, 15-16 Sept. 2005.

S.D. Naylor. "War Games Rigged?" Army Times. 16 Aug 2002. http://www.armytimes.com/legacy/new/0-292925-1060102.php

S.M. Stevens, J.C. Forsythe, R.G. Abbott, C.J. Gieseler. Experimental Assessment of Accuracy of Automated Knowledge Capture. Foundations of Augmented Cognition, HCII 2009. San Diego, CA. 2009.

S. M. Stevens, J. D. Basilico, R. G. Abbott, C. J. Gieseler, C. Forsythe. Automated Performance Assessment to Enhance Training Effectiveness. IITSEC 2010. Orlando, FL, 2010.

Stone, E.E.; Skubic, M.; "Evaluation of an inexpensive depth camera for passive in-home fall risk assessment," Pervasive Computing Technologies for Healthcare, 2011 5th International Conference on , vol., no., pp.71-77, 23-26 May 2011.

Xu, M.; Orwell, J.; Lowey, L.; Thirde, D.; "Architecture and algorithms for tracking football players with multiple cameras," Vision, Image and Signal Processing, IEE Proceedings, vol.152, no.2, pp. 232- 241, 8 April 2005.

CHAPTER 2

Augmented Navigation: A Cognitive Approach

Emrah Onal, April Savoy, Susan McDonald

SA Technologies, Inc.
Marietta, GA, USA
{emrah, april.savoy, susan.mcdonald}@satechnologies.com

ABSTRACT

User interfaces typically rely on navigation controls to help the user traverse the target system or information space. Navigation controls come in many shapes and forms, ranging from simple buttons to complex hierarchical menus for large applications. In large-scale systems, simple navigation controls may not provide sufficient cues to help the user determine where to navigate. This could be detrimental to usability and performance especially when there is penalty (e.g., as page loading time or bandwidth use) associated with inadvertent or erroneous navigation.

We propose a methodology based on cognitive modeling for augmenting navigation controls with relevant information to assist the user in negotiating the intended application space. Our methodology is based on applying a form of cognitive task analysis, goal-directed task analysis (GDTA), to determine the user's goals and decisions related to the task being performed and augmenting applicable navigation controls based on the identified situation awareness (SA) requirements. GDTA methodology has been used extensively in a variety of commercial, aviation, and military domains to determine SA requirements.

Keywords: augmented controls, navigation, live tiles, metro style, cognitive task analysis, goal-directed task analysis, situation awareness, decision-making.

1 INTRODUCTION

Navigation, specifically web navigation and application navigation, is ubiquitous in user interfaces when interacting with personal computers and mobile devices. Applications and web sites typically rely on a set of controls (e.g., buttons, menus, hyperlinks) to aid the user in traversing the system or take the user to their intended target destination. In this paper, we propose a methodology based on cognitive modeling to augment navigation controls and provide users with diagnostic information about their destination, even before navigation takes place. By providing critical supplementary information, augmented navigation informs the user about the target and helps the user decide where to navigate. Thus, augmented navigation can be helpful in reducing the number of steps when completing a task. It can be especially useful in low-bandwidth environments or slow systems, where there is a significant penalty associated with erroneous or unnecessary navigation. In addition, this approach is suitable for complex systems that present the user with a large number of potential destinations and the user has to make a series of decisions to determine the appropriate navigation target. While any type of user interface (UI) control (e.g., buttons that perform actions) can be augmented with supplementary information, for the purposes of this paper we will focus on navigation controls.

In simple terms, augmented navigation is the enhancement of UI controls with key information to assist the user in navigation decisions. Augmented controls can provide users with a valuable window into the target destination before actually navigating there. Basic examples of such controls can be found in everyday use: an email application ("app") icon may show the number of unread messages or a weather app icon may display the local temperature. Recently, Microsoft's Metro design language introduced the concept of "Live Tiles" that incorporate useful information about the underlying app ("Guidelines and checklist of tiles"). A calendar Live Tile may show upcoming events or a chat Live Tile may show the most recently received chat message. These examples point to controls that are relatively straightforward to augment. However, for complex business, military, or consumer apps, how to augment a control in a useful way may not be as apparent. Large-scale systems can have hundreds of screens necessitating complex navigation structures and associated navigation controls. Navigating such systems quickly and effectively can become a significant decision-making challenge for the user.

2   BACKGROUND AND STATE-OF-THE-ART EXAMPLES

Over the years, we have witnessed many technological advances that have aided the seamless integration of computing devices into our everyday lives – personal and professional. Throughout the varying devices and applications, there is at least one noticeable commonality: navigation controls. Productivity, financial, medical, social, or mobile apps, business and military systems, and games all have navigation controls. There is always a need for some mechanism to aid the traversal of systems, applications, or information spaces. The research and design of navigation controls date back to the first computer systems and the use of hyperlinks in early websites (Chakrabarti et al, 1998).

Navigation controls are noted as a mainstay of website design. Traditional navigation controls can be characterized as a set of links or index of options that

affords a central steering mechanism for constructed information spaces (Kalbach, J. and Bosenick, T., 2003). Successful facilitation of users' movement must achieve four key objectives:
1. Provide orientation of users' position in the system, application, or environment;
2. Support the users' determination of target destinations;
3. Enable the user to recognize and monitor the appropriate mechanisms for interaction and initiation of traversal; and
4. Supply indicators for the user to recognize successful completion of transitions (Spence, 1999).

Information presentation and information content are two aspects of navigation design essential for satisfying the aforementioned objectives. Navigation controls implement interface design as a function of what information is displayed (i.e., information content) and how that information is displayed (i.e., information presentation) (Savoy & Salvendy, 2007). These two aspects of navigation design are also used to ease the adaptation of technology by the user. As UI technologies advanced, designs have evolved producing innovative, easy-to-use, and aesthetically pleasing navigation controls, as illustrated in Figure 1.

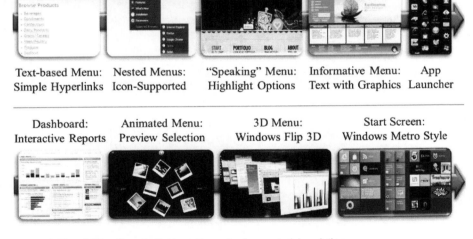

Figure 1. Pictorial timeline of milestones in navigation controls' evolution.

A large amount of research concentrates solely on information presentation (Burrell & Sodan, 2006). In addition, researchers have investigated how much information is enough to aid the user's navigation efficiency (Wei et al, 2005; Wang & Yen, 2007). However, identification of exactly *what information* to display is key in supporting the user's decision-making. The best presentation of the wrong information creates usability problems and does not aid users in accomplishing their task. The "right" information will allow the user to demonstrate optimal navigation efficiency and effectiveness.

In this paper, we will discuss a principled methodology for determining what information is needed by users to facilitate UI navigation. Our approach introduces the identification of user information requirements at the beginning of the design cycle. This approach will be helpful throughout the user-centered design process when designing low to high information-intensive navigation controls including hyperlinks, dashboards, and start screens.

3   A COGNITIVE APPROACH

In order to determine the most effective information for augmenting navigation controls, it is critical to understand the user's thinking process and the decisions they will be making. This can be done with cognitive modeling to directly support the user's situation awareness (SA). SA has been recognized as a critical foundation for successful decision-making in a variety of domains including power systems, commercial aviation (Endsley, Farley, Jones, Midkiff, & Hansman, 1998), both en route and TRACON air traffic control (Endsley & Rodgers, 1994), and the military domain (Strater, Jones, & Endsley, 2001; Strater & Endsley, 2005; Strater, Connors & Davis, 2008). SA can be described as an internal mental picture of the user's environment. Many pieces of information must be brought together to form an integrated picture of SA. This mental model forms the central organizing feature from which all decision-making and action takes place. Endsley (1988) formally defines SA as "...the perception of the elements in the environment within a volume of time and space, the comprehension of their meaning and the projection of their status in the near future." The definition encompasses several concepts that are important in understanding the SA construct. SA is comprised of three levels: perception, comprehension, and projection (see Figure 2).

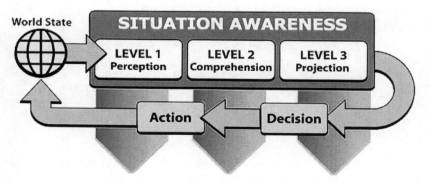

Figure 2. SA is a key component for successful decision-making. Building good SA starts with giving users the right information supporting level 1 (perceiving basic data), level 2 (synthesis and interpretation), and level 3 (anticipating future states) SA formation (Endsley & Jones, 2012).

Level 1 SA, perception, involves the detection of significant environmental cues. Level 2 SA, comprehension, involves integrating and comprehending the information to understand how it will impact the individual's goals and objectives. Level 3 SA involves extrapolating this information forward in time to determine how it will affect future states of the operating environment (Endsley, 1988; Endsley, 1993).

We can formally define augmented navigation as the practice of embedding informational cues directly into the navigational structure of user interfaces for the purpose of: a) aiding the user in deciding where to navigate, and b) supporting high levels of SA for decision-making. There are several well-established design methodologies that may be used in determining how best to augment navigation controls with relevant information. For example, user-centered design, interaction design, and goal-oriented design methodologies all aim to aid the design process by utilizing personas (i.e., user roles), scenarios, and user goals. While all of these design methodologies implicitly support SA by using best design practices, none of them do so in an explicit manner. In order to support SA in a structured way, we propose goal-directed task analysis for requirements analysis that is part of the SA-oriented design methodology documented by Endsley & Jones (2012).

3.1   Goal Directed Task Analysis

Goal-directed task analysis (GDTA) is a form of cognitive task analysis typically used for SA requirements analyses (Endsley, 1993; Endsley & Jones, 2012). The GDTA involves knowledge elicitation with subject matter experts (SMEs) and users in order to identify the goals of a particular user role and to define the sub-goals for meeting each higher goal. Associated with each sub-goal are the critical decisions that must be made in order to meet the user's objective. For each decision, the SA requirements are delineated. SA requirements are the essential the pieces of information needed to make the decision. This goal-oriented approach focuses on the ideal cognitive requirements for reaching a goal, instead of on basic task steps. GDTA methodology has been used extensively to determine SA requirements in a wide variety of operations including power systems, commercial aviation (Endsley, Farley, Jones, Midkiff, & Hansman, 1998), both en route and TRACON air traffic control (Endsley & Rodgers, 1994), and the military domain (Strater, Jones, & Endsley, 2001; Strater & Endsley, 2005; Strater, Connors & Davis, 2008).

Figure 3 shows a conceptual GDTA for a user responsible for the mission plan of a hypothetical Mars exploration rover. This hierarchy shows the user's goals and sub-goals, including the related decisions and information requirements. The decisions are presented as questions that must be addressed in order to meet the goals. SA requirements for "Decision B" are shown in the box titled "Information Requirements", and include the perceptions (level 1 SA -- innermost indentation), comprehensions (level 2 SA -- mid-level indentation), and projections (level 3 SA -- outermost indentation) that support the decision-making process. Such a GDTA provides a model of the cognitive requirements for the domain. A thorough SA requirements analysis for a user's role serves as the foundation for UI design, including the augmentation of navigation controls with the right information.

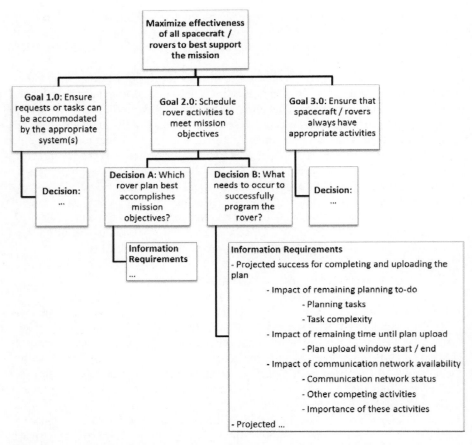

Figure 3. Hypothetical GDTA snippet for communicating with Mars exploration rovers. The information requirements shown here will be used to demonstrate an augmented navigation control. *Note: This limited GDTA was created for the purposes of this paper to demonstrate how GDTA can be used to augment navigation, and does not take into account the many factors, considerations, and activities associated with planning and executing a rover mission.*

3.2 Augmenting Navigation

Figure 4 shows sample navigation controls for the Mars rover operation described in the previous section. "Communication Controls" is a hypothetical screen for Mars exploration rovers; specifically for managing communications. Figure 4 (a) is a typical navigation button showing the name of the target screen with no supplemental information. Figure 4 (b) is "augmented" with basic additional information. Augmentations in Figure 4 (c) are based on the information requirements outlined in Figure 3 (indentation levels correspond to levels of SA), showing level 2 SA. Focusing on level 2 (e.g., "Com. window in 60 mins") and level

3 (projections or estimations on future system states) is recommended since higher levels provide processed information that reduces the user's cognitive workload. Note that a good system design should also provide the means to get to level 1 information (e.g., "Com. window at 04:00 GMT") when necessary for the user. Clearly, Figure 4 (c) provides more relevant information compared to (a), supporting the user's decision-making process.

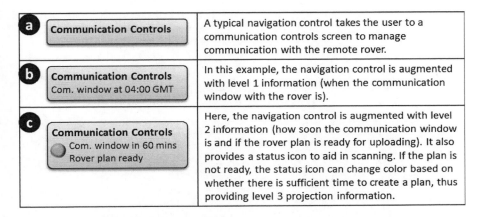

Figure 4. Navigation buttons in (a), (b), and (c) show no augmentation, basic augmentation, and level 2 SA augmentation, respectively.

3.3 Degrees of Augmentation

The hierarchical structure of GDTA allows for varying degrees of augmentation. More gross levels of augmentation apply to higher-level, overarching goals. For example, Figure 5 (a) shows "alert" buttons in a hypothetical Navy anti-submarine warfare (ASW) system. These buttons show the number of unacknowledged and acknowledged warnings (signified by a red octagon with an exclamation mark) and cautions (signified by a yellow triangle with an exclamation mark) that are currently active. New unacknowledged alerts result in a colored "swipe" across the button to draw the user's attention. Selecting these buttons would navigate to the alerts screen. In almost any military situation, the highest level goal is to complete the mission successfully (whatever that mission might be for a given user). Because properly defined alerts always pertain to events that threaten life or mission success, this minimal level of augmentation is relevant to the highest level goal of successful completion of the mission ("Execute ASW Plan" goal in the hypothetical example provided in Figure 6). Figure 5 (b) shows a more detailed augmentation mechanism relevant to a lower level goal. In this example, information about the most current warning and caution is provided. Selecting the icon would take the user to the screen relevant to that alert. In this case, the "unknown ship" alert is relevant to the goal "Establish contact location and classification" in Figure 6, which is accomplished on another screen, so selecting this alert would navigate directly to the relevant screen. Finally, for very specific goals, particularly those related to information-dense domains, more elaborate augmentation may be beneficial. Figure 5 (c) shows a small window that provides high-level information on the status of each ship in a group with respect to a given search plan. If any ship's deviation from the plan reaches a critical threshold, the small window provides the user with awareness of the delta, and the elongated button along the edge of the window can be used to immediately navigate to the relevant screen where the user can address the problem. Deviations from the current plan are part of the goal "Monitor Adherence of Friendly Assets to the Plan" in Figure 6.

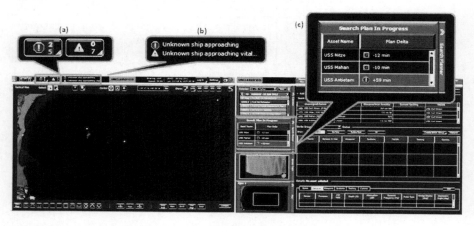

Figure 5. Examples in (a), (b), and (c) show different degrees of augmentation.

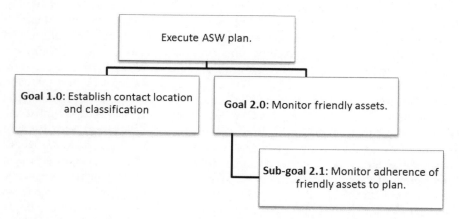

Figure 6. Hypothetical GDTA snippet for anti-submarine warfare (ASW). Only top level goals are shown.

3.4 Considerations

There are many considerations to take into account when augmenting navigation controls to support SA-oriented system navigation. The target medium can influence navigation interactions and information visualization, and may constrain how controls can be augmented. For example, mobile devices bring unique constraints based on screen size, battery reserves, and available bandwidth. Designs using the augmentation methodology are affected by several factors:
- Screen size and resolution -- added information and visualizations increase the footprint of the control.
- Bandwidth availability, data usage considerations, battery life, and processor use may impact the amount of information shown on the navigation control and its update frequency.
- The tradeoff between system resources used to update the supplementary information versus the cost of erroneous navigation and user's lack of SA.
- Navigation complexity and the need for navigation related decision-support in the application.

4 DISCUSSION

We believe that SA-oriented system navigation, or augmented navigation, can be effectively designed and specified using the SA-oriented design methodology. This approach is suitable for a variety of domains and from simple to complex applications where navigation support is needed. Furthermore, we believe that by directly supporting high levels of SA, augmented navigation can help the user's decision-making and increase usability.

While we believe that SA theory firmly supports the design and use of augmented navigation, this theory has to be objectively demonstrated by empirical studies showing the advantage -- or disadvantage -- of using augmented navigation controls. Our team is using this methodology in a number of projects, and in the future we hope to report concrete findings.

REFERENCES

Burrell, A. & Sodan, A. (2006). Web interface navigation design: Which style of navigation-link menus do users prefer? Paper presented at the 22nd International Confernece on Data Engineering.

Chakrabarti, S., Dom, B. & Indyk, P. (1998). Enhanced hypertext categorization using hyperlinks. In Proceedings of the 1998 ACM SIGMOD international conference on Management of data (SIGMOD '98), Ashutosh Tiwary and Michael Franklin (Eds.). ACM, New York, NY, USA, 307-318.

Endsley, M. R. (1988). Design and evaluation for situation awareness. In Proceedings of the Human Factors Society 32nd Annual Meeting (pp. 97-101). Santa Monica, CA: Human Factors and Ergonomics Society.

Endsley, M. R. (1993). A survey of situation awareness requirements in air-to-air combat fighters. *International Journal of Aviation Psychology*, **3**(2), 157-168.

Endsley, M. R., & Jones, D. G. (2012). *Designing for Situation Awareness: An Approach to User-Centered Design.* (London: Taylor & Francis).

Endsley, M. R., & Rodgers, M. D. (1994). Situation awareness information requirements for en route air traffic control (DOT/FAA/AM-94/27). Washington, DC: Federal Aviation Administration Office of Aviation Medicine.

Endsley, M. R., Farley, T. C., Jones, W. M., Midkiff, A. H., & Hansman, R. J. (1998). Situation awareness information requirements for commercial airline pilots (ICAT-98-1). Cambridge, MA: Massachusetts Institute of Technology International Center for Air Transportation.

Garrett, J. J. (2003). The elements of user experience: User-centered design for the web. New Riders.

"Guidelines and checklist for tiles," Accessed February 23, 2012, http://msdn.microsoft.com/library/windows/apps/hh465403.aspx.

Kalbach, J., & Bosenick, T. (2003). Web page layout: A comparison between left and right-justified navigation menus. *Journal of Digital Information,* **4**(1).

Savoy, A. & Salvendy, G. (2007). Effectiveness of content preparation in information technology operations: Synopsis of a working paper. In Proceedings of the 12th International Conference on Human-Computer Interaction, pp. 624-631.

Spence, R. (1999). A framework for navigation. *International Journal of Human-Computer Studies*, **51**, 919-945.

Strater, L. D., & Endsley, M. R. (2005). Designing to enhance SA in the CIC. In Proceedings of the Human Systems Integration Symposium 2005. Washington, DC: American Society of Naval Engineers. Published on CD-ROM.

Strater, L. D., Jones, D. G., & Endsley, M. R. (2001). Analysis of infantry situation awareness training requirements (No. SATech 01-15). Marietta, GA: SA Technologies.

Strater, L. D., Connors, E. S., & Davis, F. (2008). Collaborative control of heterogeneous unmanned vehicles. Paper presented at the Texas Human Factors and Ergonomics Conference.

Wang, M. and Yen, B. (2007). Web structure reorganization to improve web navigation efficiency. Paper presented at the 11th Pacific-Asia Conference on Information Systems.

Wei, C., Evans, M., Eliot, M., Barrick, J., Maust, B., and Spyridakis, J. (2005). Influencing web-browsing behavior with intriguing and informative hyperlink wording. *Journal of information Science,* **31**(5), 433-445.

CHAPTER 3

Change Blindness When Viewing Web Pages

Daniel Steffner, Bo Schenkman

Stockholm University, Stockholm, Sweden
Blekinge Institute of Technology, Karlskrona, Sweden and Speech, Hearing and Music, Royal Institute of Technology (KTH), Stockholm, Sweden
Email: bo.schenkman@bth.se

ABSTRACT

Change blindness on web pages was studied for 20 participants. The purpose was to find how change blindness appears for web pages, and which changes are easier to detect. The task was to detect if a change had occurred and to show this by the means of the cursor. Rensink´s flicker paradigm was used, where four categories of changes were presented. It was easier to detect a change not consisting of a person than one with a person. It was easier to detect a change to the left than to the right. The complexity of the web pages did not appear to have an effect, while large changes were easier to detect than small. The results may indicate that focused attention is differently sensitive for different kinds of changes. They also show that change blindness is a general phenomenon that can be applied to the perception of web pages.

Keywords: Change blindness, attention, perception, web pages, reaction time

1 INTRODUCTION

Change blindness is the phenomenon when a person does not detect that something in the visual field has changed. These changes may be large or minor. Most research has to a large extent focused on how this phenomenon appears in natural situations or when a person looks at a visual scene, e.g. a photo. One of the early studies was done by French (1953), who used a discrimination paradigm for these effects. Among the various developments when studying change blindness was the introduction of the flicker paradigm by Rensink, O'Regan and Clark (1997) where two scenes are presented in succession, with a blank picture, a mask, put between them for a short period, e.g. 80 msec. During this short interval large changes may be made in the second picture, without the observer noticing these. An original picture, P, is shown. It is then followed by a mask which is a gray image in order to hide detection of motion. After the short presentation of the mask, the changed picture, P' is shown, see Figure 1. The whole sequence is then repeated till the change is detected. There is a need for attention demanding mechanisms in order to see the change (Rensink, 2002). If the mask was not present the change would be easily detected because of the perceived motion.

A common explanation of change blindness is based on how the visual short term memory functions. Visual information is believed to be transmitted through a filter bottleneck demanding attention to a structure for short visual storage with low capacity. In order for the visual information to be processed to acknowledge a change, there is a demand for a constant viewing of what is viewed (O'Regan, 2007). Rensink (2000, 2002) has proposed the coherence theory that is in accordance with the perceptual characteristics of change blindness. Simons (2000) has proposed five hypothetical causes as explanation for change blindness, e.g. overwriting of older information. However, it may be that change blindness is not due to the limited capacity of the visual short term memory, but to a failure to engage it (Triesch et al, 2003). Simon's causes would then become effects of change blindness.

It is important to distinguish between inattentional blindness and change blindness, see Jensen et al (Jensen, Yao and Simons, 2011). Inattentional blindness is the failure to notice the existence of an unexpected item, while change blindness is the failure to notice an obvious change. They have similar phenomenologies, viz. a person does not see something obvious, but their etiologies and theoretical implications for vision are different. Chabris and Simons (2010) have written an entertaining book, where incidents of inattentional blindness are interpreted as a form of everyday illusion. In this paper, however, the focus is on change blindness.

In recent studies on change detection, one has made use of presentations or displays closer to real world perception (Simons, 2000). Besides using more or less artificial stimuli in laboratory conditions, one may also study change blindness in situations in real life, in natural images such as photographic pictures or in role games arranged by a test leader. According to Simons and Rensink (2005), the counterintuitive findings of change blindness in naturalistic conditions strongly support the argument that it is not an artifact, but some general failure to retain or compare information from moment to moment. The generality of the phenomenon is also suggested, when one notes that change blindness may occur in other sense modalities than vision. In hearing it is called change deafness. See e.g. Fenn et al (2011), who studied change deafness in the context of telephone conversations.

One new interaction tool is the Internet with its millions of web pages. Web pages change, some slowly, while others change rapidly. It is therefore of practical and ergonomic interest to see how changes on web pages are detected. Secondly, since a web page constitutes a non-natural artifact, it is of interest to understand how a person perceives and detects changes for such objects. A review of how both inattentional blindness and change detection affect human-computer interface design issues may be found in an article by Varakin, Levin and Fidler (2004). People are unaware of their change blindness, which Varakin et al call Change Blindness Blindness. People overestimate their visual processing capacity, that is constituted by illusions of the observer. The consequences may be seen in less than satisfactory user interfaces. We therefore found it important to understand change detection when a person is in contact with the Internet and web pages.

This study had as it aims (1) to investigate change blindness in the context of web pages, when an observer uses web pages, and (2) how different kinds of changes affect their detection.

We chose to study four kinds of changes. (1) Size: Is it more difficult to detect big changes than small ones? (2) Complexity: Is the complexity of the content a factor that will affect the detection? (3) Location: Is the location of the change important for detection. Schenkman and Fukuda (2001) found e.g. that observers of web pages preferred objects to the left. (4) Person: Are we more attentive to changes of a person than to a non-person. Our hypotheses were the following: (1) large changes, (2) changes of simple objects, (3) changes to the left and, (4) changes relating to a person, would all be easier to detect than their opposites.

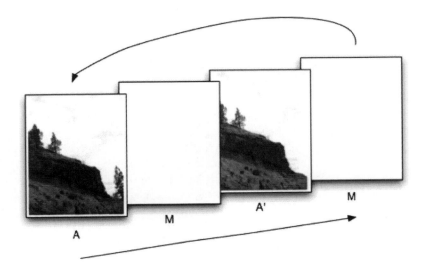

Figure 1 Rensink's flicker paradigm. A mask (M) is shown between the images (A) and (A'). The mask effectively removes the movement normally associated with change. In this case between the image (A) and the altered image (A').

2 METHOD

2.1 Participants

The flicker method by Rensink (2002) was used to present different web pages to 20 test persons, all university students, aged 19 to 42 years, mean = 27.5 years. All except one person reported that they had good vision, either uncorrected or when corrected with glasses or lenses. The one person who reported that she did not have good vision did, however, use neither glasses nor lenses. One requirement for participation was that they had had been using computers, which all of them had been doing for 4 to 12 years.

2.2 Experimental design

The design of the study was a within-design, with one category variable, where the levels of this variable were the different web pages.

The independent variable was the different web pages, where different types of changes, viz. size, complexity, location and person, had been introduced. During up to 60 seconds each pair with blank page, the mask, between them was presented for 240 ms + 80 ms + 240 ms. There were two dependent variables. One variable was the detection time for noticing the change. The second dependent variable was the locus of the change that the person had to mark with the cursor.

2.3 Web pages

The test program was constituted of 28 computer based presentations or slideshows, where 24 had been designed according to the flicker paradigm by Rensink, O'Regan and Clark (1997), as presented in Figure 1. In four of these presentations there was no change in any picture. The 24 sequences were divided into four categories of web pages as to the four hypotheses we wanted to test, thus six sequences in each category. The general layout of the 24 different slide-shows is shown in Figure 2. Of these six, three of them were made to belong to one of the tested sub-categories, while the other three belonged to the other sub-category. Thus, for the category 'complexity', three of the sequences had changes on simple web pages, while three had changes on complex web pages.

The 'simple' web pages had five or fewer web page elements that were distinctly separated with clearly defined borders and margins. The 'complex' web pages had at least nine web page elements that were placed much closer and without distinct separators. For 'location', web page elements with changes were placed either to the left, or right hand side of the web pages.

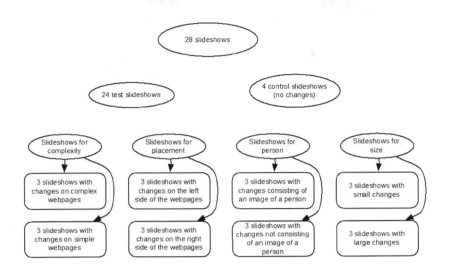

Figure 2. Division of slideshows into categories and sub categories. 28 slideshows were made, 24 for tests and 4 for control. The 24 test slideshows were divided into four categories for the four selected kinds of change. Each of these categories were then divided into two opposing sub categories, each consisting of three different slideshows.

An example of how a slideshow could look is shown in Figure, where a pair form the image category Large is shown. The differing feature of the two images is seen in the lower right part.

Figure 3. The composition of a slideshow, without the masks between the two images. The image on the left is unchanged, while the one on the right has been changed.

By considering the web pages as divided into a grid of three rows and three columns, nine equally sized quadrants of placement were used to position the web page elements with changes. The three web pages with changes on the left side each had their web page element with the change in one of the three quadrants in the left column. The analogous principle applied for changes on the right side. For 'person', three had changes on an image of a person, while three classified as 'non-person' consisted of an image of an object or a block of text. For 'size', the changed web page elements categorized as "large" were at least five times the size of those categorized as "small". The presentations were shown to each participant in a random order.

2.4 Room conditions and equipment

The image sequences were presented on a 21 inch CRT screen with a resolution of 1024x768, using a picture frequency of 100 Hz. Luminance at the center of the screen with maximum contrast and lightness was 170 cd/m2, when the room illuminance in the horizontal plane was kept at 120 lux, when measured in front of the screen.

2.5 Procedure

The participants were informed on paper that they would see a computer presentation with flicker. In some of these there would be a change in the picture. Their task was to detect this eventual change and to mark with the cursor where this change had taken place. They had a time period of 60 s to detect the change. If they did not succeed within this time duration, the sequence was stopped, and the person was informed if there had been a change that went undetected, or whether there had been no change in that particular sequence. The detection time for the person for each sequence was recorded as was the location where he/she had marked the eventual change.

Before the actual experiment, the participants had a training session of four sequences in order to familiarize them with the equipment, task and procedure. After half of the trials, i.e. after 12 sequences, a short break was had. The total time for completing the experiment varied between the participants, but none lasted longer than 40 minutes.

3 RESULTS

3.1 Detection times

The detection times for the different slideshows are shown in Figure 4. The mean detection times varied much between the different pictures. On average over all the participants the detection time was 19.6 s, SD = 4.2 s. The fastest individual had a mean detection time of 12.7 s and the lowest had a mean time of 25.9 s. One-way analysis of variance showed that the changes in the pictures were detected differently fast. Sphericity was not obtained, as shown by Mauchly´s test, and the degrees of freedom were therefore corrected. Huyn-Feldt showed significance, $F(4,462) = 15.6$, $p < 0.001$, as well as the Greenhouse-Geisser test, $F(4,872) = 15.6$, $p < 0.001$. Reaction times were not logarithmically transformed, since visual inspection of the data indicated that they appeared to be approximately normally distributed.

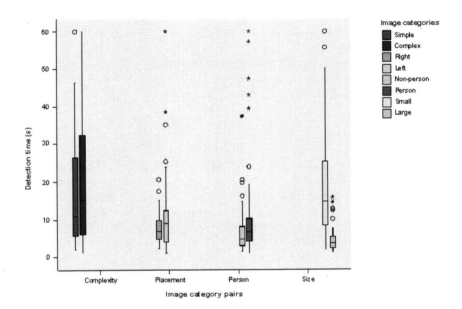

Figure 4. Detection times for the categories and their constituent sub-pairs. Boxes represents standard deviations. o represents outliers, with values within 1.5 and 3 box lengths from the upper quadrille and * represents extreme values with values more than 3 box lengths from the upper quadrille.

For thirteen of the presentations the changes were detected by all participants. In some cases, the observer did not detect the change at all. For five presentations and for another five, one respectively two persons failed to detect the change. For one picture, four people did not detect the change. There were no false positives.

The participants were slowest to detect change in one of the images that had been categorized as simple, constituted of a web page with few objects and with a small change. They were fastest, i.e. having the lowest mean detection time, for a change in a web page having a large change.

Post-hoc analysis of pairwise comparisons with a significance level of α equal to 0.05 showed that changes to the right of the web page, of a non-person object and of big objects were significantly easier to detect than objects on the left, of an image of a person and of small objects, respectively. There was no significant mean difference between web pages that were considered as having simple or complex variations.

3.2 Search strategies

The participants had also been asked about their search strategies. The fastest person, 20 years old, had first looked on the whole picture, and if not detecting the change had begun to look systematically from left to right. Ten of the others said that they first had tried to perceive the whole page and then had begun to search in more detail. Eight of the participants did not experience the tests as difficult, which six of them did. Interestingly enough, three participants said the tests were strenuous for the eyes and one of them experienced a slight nausea.

4 DISCUSSION

There are a number of factors that may affect the speed of change detection, as listed by Durlach (2004), where she e.g. mentioned distractions, discriminability and categorization. She also discussed methods to facilitate change detection in visual displays, especially regarding complex system design and operator training. However, for some applications like web pages, it may be that the designer desires a certain degree of change blindness. Such a view was presented by Intille (2002). He proposed that for ubiquitous computing, a designer should use change blindness as a tool. The user would then not see changes and "calmness" would be preserved.

Different kinds of changes on web pages affect how easily they are detected. A little surprising in our study was that changes concerning persons were less easily detected than non-person objects. It is possible that the participants put less weight at images of persons than at other objects. This explanation is supported by findings of Xu (2005), who in studies on face recognition concluded that we are not more predisposed to detect faces than other kinds of objects. However, Ro, Russell and Lavie (2001) reached the opposite conclusion. Faces were more easily detected when participants were subjected to the flicker paradigm. It could be that in some situations, observers are more attentive to faces. This could be due to cultural effects (Masuda and Nisbett, 2006).

The lack of significance between simple and complex web pages could be caused by an insufficient difference in complexity. It may be that there is a high tolerance to visual clutter before it hinders change detection.

Another unexpected result in the present study is that objects to the left on the screen were more difficult to detect than objects to the right. This is in contrast to what Schenkman and Fukuda (2001) found on aesthetic preferences for web pages containing text. This could mean that people do not perceive web pages with images in the same way as we perceive web pages with text that we are supposed to read. When changes occur on a web page that has been reloaded by a user, the change may be difficult to detect. Sometimes it might be in the interest of the user that it is emphasized that a change has occurred. In other circumstances, the opposite might apply. People often think that they will notice important events, which reflect a misunderstanding of how attention functions, see Simons and Ambinder (2005). If a person, for example, is to watch a warning display, change blindness may make him/her fail to detect the change. If the change is not within the focus of attention, it may go unnoticed.

The results may also indicate that we do not view web pages in the same manner as we view printed text. The study also gave further evidence that change blindness is a general phenomenon which may occur in many various situations or for various kinds of presentations. However, Simons and Rensink (2005) warn for drawing too strong conclusions from studies on change detection to how the visual processes function. For example, a failure to detect a change does not imply the absence of a (mental or visual) representation. Simons and Rensink argue that change blindness studies can only reveal limits of the representation for the conscious perception of dynamic change. It is also noteworthy to observe that change blindness to some extent is culturally dependent (Masuda and Nisbett, 2006), who found that East Asians had their attention on context, while Americans had theirs on focal object information, when observing images with change blindness. The present study was conducted with Swedish participants, and they may differ from both Americans and East Asians. Culture must thus be considered when using change blindness for web applications. We believe that the present results should be of relevance for web designers and designers of user interfaces.

REFERENCES

Chabris, C. and D. Simons. 2010. *The invisible gorilla: And other ways our senses deceive us.* New York: Crown Publishers.

Durlach, P.J. 2004. Change blindness and its implications for complex monitoring and control systems design and operator training. *Human-Computer Interaction* 19: 423-451.

Fenn, K.M., H. Shintel, A.S. Atkins, J.I. Skipper, V.C. Bond and H.C. Nussbaum. 2011. When less is heard than meets the ear: Change deafness in a telephone conversation. *The Quarterly Journal of Experimental Psychology* 64: 1442-1456.

French, R.S, 1953. The discrimination of dot patterns as a function of number and average separation of dots. *Journal of Experimental Psychology* 26, 1-19.

Intille, S.S. 2002. Change blind information display for ubiquitous computing environments, In *UbiComp 2002*, LNCS 2498, eds G. Borriello and L.E. Holmquist. Berlin Heidelberg: Springer: 91-106.

Jensen, M.S., R. Yao, W.N. Street, and D.J. Simons, 2011. Change blindness and inattentional blindness, *WIREs Cognitive Science* 2: 529-546.

Masuda, T. and R.E. Nisbett, 2006. Culture and change blindness. *Cognitive Science* 30:381-399.

O'Regan, J.K. 2007. Change Blindness. Accessed 23 May, 2011, http://nivea.psycho.univ-paris5.fr/ECS/ECS-CB.html

Rensink, R.A. 2000. Seeing, sensing and scrutinizing. *Vision Research* 40: 1469-1487.

Rensink, R.A. 2002. Change detection. *Annual Review of Psychology* 53: 245-277.

Rensink, R.A, J.K. O'Regan, and J.J. Clark. 1997. To see or not to see: The need for attention to perceive changes in scenes. *Psychological Science* 8: 368-373.

Ro, T., C. Russell, and N. Lavie. 2001. Changing faces: A detection advantage in the flicker paradigm. *Psychological Science* 12: 94-99.

Simons, D.J. 2000. Current approaches to change blindness. *Visual Cognition* 7: 1-15.

Simons, D.J. and M.S. Ambinder. 2005. Change Blindness: Theory and consequences. *Current directions in Psychological Research* 14: 44-48.

Simons, D.J. and R.A. Rensink. 2005. Change blindness: Past, present and future. *Trends in Cognitive Sciences* 9: 16-20.

Schenkman, B. and T. Fukuda. 2001. Aesthetic appreciation of web-pages as determined by category scales, magnitude estimations and eye movements. In *Proceedings of the International Conference on Affective Human Factors Design*, eds M.G. Helander, M.K. Halimathun and T.M. Po. London: ASEAN Academic Press: 309 – 316.

Triesch, J., D.H. Ballard, M.M. Hayhoe and Sullivan, B.T. 2003. What you see is what you need. *Journal of vision* 3: 86-94.

Varakin, D.A., D. T. Levin and R. Fidler. 2004. Unseen and Unaware: Implications of Recent Research on Failures of Visual Awareness for Human–Computer Interface Design, *Human-Computer Interaction* 19: 389-342.

CHAPTER 4

Qualitative Facets of the Problem Statement

von Brevern, Hansjörg; Karwowski, Waldemar

Independent Consultant; University of Central Florida
Zürich, Switzerland; Orlando, USA
vonbrevern@acm.org; wkahfe@gmail.com

ABSTRACT

One central reason for today's ill-specified man-machine systems at the forefront are ambiguous, incomplete, grammatically incorrect, and isolated written descriptions of features and requirements as well as the absence of workflows and processes of human task strategies. We therefore discuss affects from a larger context during elicitation, analysis, negotiation, decision-making, validation, design, and realization from the eyes of human involvement and task performance. We present the root of feature and requirements lists that still prevail in today's organizations and contrast descriptions of features and requirements with those of a problem statement. Under the umbrella of the OO paradigm, we illustrate dependencies of narratives within the problem domain with its interdependent models and views. As a suitable method for an organization for less ambiguous descriptions of deltas and behaviors, we present the heuristics-based method 'class, responsibilities, and collaboration' (CRC), present practical guidelines, and give examples, in which human mental and motor actions may need to be included.

Keywords: Problem statement; task description; requirements description; requirements engineering; object-oriented analysis and design (OOAD); class, responsibilities, and collaboration (CRC)

1   THE LARGER CONTEXT

(DeMarco, 1996) observed, "ill-specified [man-machine] systems are as common today as they were when we first began to talk about Requirements Engineering". In 2012, this holds still true for informally written features, mutually dependent requirements, processes, task strategies, and the like. One major issue is the descriptive manner in view of their quality of style, granularity, depth, who's, what's, when's, and how's as a result from elicitation, analysis, negotiation, and decision-making between human subjects.

Too often, too many 'irrelevant' and unaligned motives, goals, tasks, and interests of different stakeholders, their levels of intellectual thinking, evaluations, and sensory-perceptual reflections of situations, budget, given time constraints, and even project management methods stimulate an unfortunate culture of fairly vague descriptions. Regrettably, project management methods often "attempt to inappropriately constrain solid engineering where they should not" (von Brevern and Synytsya, In press). Nevertheless, the new delta or requirement is a shared agreement, 'written contract', and working assumption between business and IT stakeholders. It is not uncommon in organizations that those human subjects who actually use the artifact as their daily object of goal-oriented work activity towards a result of activity are incorrectly excluded from elicitation, analysis, negotiation, and decision-making requirements engineering processes. Overall, praxis proves that ambiguous descriptions of requirements foster gaps between expected behaviors by human subjects, systems, and business, can lead to errors or even fatal failures, cause delays, affect human work performance, motivation, and productivity, increase costs, and more. Hence, an unambiguous and well-structured verbalization of a meaningful and situated problem, delta, or requirement that must satisfy different levels of abstractions, degrees of granularity, and viewpoints towards a solution delivery is extremely challenging. Although foci of this paper are not on negotiation, decision-making, or dialectic inquiry, newly agreed deltas can affect human goal-oriented work activity, human motivation, productivity, and task performance, operational and system processes, system design and development, organizational efficiency, competitiveness, and costs. Therefore, new deltas or requirements must be precise and accurate prior to decision-making because of their immediate intangible and tangible impacts onto human goal-oriented work activity, processes, the environment, and the organization once agreed.

Software engineering differentiates between the problem and the solution domain. A delta or requirement is primarily a statement of the problem or subject domain although it overlaps with the solution domain in terms of analysis and states of the static structure. (Bedny and Karwowski, 2007, p. 308) argue that "[c]omputerization of man–machine systems increases the role of intellectual components of work". In this sense, the unity of human cognition and behavior, evolution of the human subject using the artifact, human learning, and skills challenge the specification of a delta. In contrast with common practice, these facets from human-centered viewpoints further support our argument for precise, complete, and unambiguous descriptions towards accuracy and reliability. As such, "[a]ccuracy characterizes the precision with which the goal of task is achieved. Reliability refers to failures of performance and how probability of failure can change over time or in stressful situations… human performance can be precise but not reliable, and not all errors can be considered as system failures" (Bedny et al., 2010, p. 378). To build reliable systems, the inclusion and description of human performance is therefore a critical part of the problem domain. From a functional view, a problem portrayed by the delta must contain the "structure as well as its user-observable properties, functional and non-functional alike. By removing the counterproductive boundary between requirements and design, a holistic view of product conception emerges" (Taylor and van der Hoek, 2007).

Those who must be able to understand the delta that enunciates the problem analyze and design potential man-machine system solutions and use engineering methods. In other words, all stakeholders – including decision-makers – should be able to proficiently communicate in this language.

2 CHALLENGES FROM CONVENTIONAL ENGINEERING

In the 1970's, procedural-programming languages supported the structured approach. Today, the structured approach still dominates many organizations and only focuses on functional views with different models at various stages throughout the software development process. Typically, the structured approach asks for feature and requirements lists and traceability matrices to exemplify system needs. Contrary to a problem statement that entails the universe of a problem domain, a single feature and requirement is a context-*independent* static snapshot. Quite often, such a snapshot is only one sentence that is hard to understand, contains meaningless '...' or 'etc.', is unspecific, grammatically incorrect, incomplete, and disregards boundaries of the universe within its environment. Furthermore, each single change of a feature or requirement not only urges a careful requirements management process but also may even negatively impact other layers of the conventional non-iterative waterfall model. Likewise, changing features and requirements can put system development at stake. Additionally, subjective interpretation of a context independent single feature or requirement may outweigh other features or requirements i.e., foci and the 'whole picture' drift even more out of sight. Therefore, context-independent islands of features or requirements together necessitate so-called requirements traceability matrices. Particularly in terms of large systems, rapidly growing feature and requirements lists and traceability matrices become unmanageable, unfocused, and unreadable. Moreover, procedural analysis and design, its feature and requirements lists, and traceability matrices are mainly system-solution-oriented. Often, organizations handle feature and requirements lists in Excel or PowerPoint slides while traceability matrices are often just a tool for project management.

Generally, neither feature and requirements lists nor traceability matrices can reflect an entire universe from human-centered perspectives because the human "thinking process reconstructs reality into a dynamic model of the situation" (Bedny et al., 2003, p. 447). So, the "… pitfall of attempting to resolve human activity by conventional engineering is that it attempts to model systems on "stimulus – response" behaviours. This approach largely ignores cognitive mechanisms of humans and with them the embedded roles, context, tasks, and goals of technical systems. Instead, the answer of how to envision, understand, and model technical systems should be explored in human activity and its manifestation" (von Brevern and Synytsya, 2006, p. 102). Human goal-oriented activity is a hierarchically ordered, individual, and unified system with organic, psychological, and socio-psychological features (Bedny and Karwowski, 2007). It goes without saying that a feature or requirement that is expressed in one sentence neither can nor is able to reflect any of the extent discussed.

More practically, **Table 1** contrasts fictional examples from a feature and requirements list as well as *extracts* from a problem statement. #1 of **Table 1** illustrates that a feature and requirement can be derived from the extract of a problem statement, which, on the contrary, cannot be done from a feature or functional requirement as shown in #3 of **Table 1**. From the extract from a problem statement of #2 of **Table 1** neither a feature nor a requirement can be deduced from the angle of the structured approach. From OO perspectives, this description allows to deduce a hierarchical structure, dependencies, and classes. Contrary to conventional descriptions of features/requirements, the examples of the problem statements in #1 and #2 reflect and narrate reality.

Table 1 Comparisons between features/requirements & a problem statement

#	Feature / functional requirement	Extracts of a problem statement
1	The ability to generate reports.	The store manager will be able at any time to print a summary report of sales in the store for a given period, including assignment of sales to sales assistants to calculate weekly sales bonuses...
2	Cannot be derived	A league is a group of teams that compete against each other. Each team recruits members to participate in the contests...
3	Supports basic internet based purchasing functionality.	Cannot be derived

In summary, **Table 1** illustrates shortcomings of conventional descriptions of feature or requirements vis-à-vis the problem statement, which, on the contrary, can be applied in OO and the legacy functional approach alike.

3 THE PROBLEM DOMAIN IN SITU

The goal of a problem domain in which usually human subjects interact with man-machine systems is to describe needs that arise from human goal-oriented work activity. From mere engineering views, the union of the problem domain, messages that cross its interfaces, inclusions, and exclusions must be identifiable by the system. With the emergence of the OO paradigm, it eliminates major weaknesses of the structured approach because it models software as a collection of collaborating objects that are part of consistent models. Contrary to the functional approach, OOA aims to primarily understand what the system will need to do, but not how it will do it until later in the design stage i.e., OOD.

One major objective of domain analysis is the construction and availability of a class model for reuse and sharing with intersecting subordinate applications. Therefore, the problem domain is based on business, (sub) processes, task strategy/performance workflows, and the problem statement (c.f., Figure 1 and Figure 2). Notably, these types of workflows *and* the problem statement coexist. Per se, the problem statement therefore describes needs at domain level as opposed to subordinate application levels within a domain.

Foci of the problem statement are on objects, their relationships, responsibilities, and collaborations within the problem domain rather than on procedures. Doing so eliminates isolation because "[o]ne of the distinguishing features of object design is that no object is an island. All objects stand in relationship to others, on whom they rely for services and control" (Beck and Cunningham, 1989, p. 2). In conclusion, with a problem statement in situ, requirements are embedded into the 'larger' whole of the universe of the subject domain.

4 THE PROBLEM STATEMENT IN SITU

Organizational practice unfortunately often remains ignorant of the ideal symmetry between the strengths of a problem statement and OOAD. The boundary of the problem domain determines dependent models like the use case model, behavioral model, and the domain class model (c.f., Figure 1). Consequently, use case descriptions and textual analyses for static structure modeling depend on the problem statement of the problem domain as illustrated in Figure 1.

A problem statement consists of at least one or more concise problems written in the jargon of the problem domain and a summary of circumstances. The jargon used is subject to a shared and commonly agreed domain dictionary, which is parent of an application dictionary that may use data dictionaries, which per se must be coherent with the domain dictionary. Problem statements need to include the roles of their human subjects, goal(s) of activity, describe task strategies, task-goal as well as subsequent descriptions of human mental and motor actions in time and space in a hierarchical order, and system-identifiable tools vis-à-vis expected results of activity. Likewise, problem domain relevant stimuli from e.g., human sensory, simultaneous-perceptual, deductive, mnemonic, creative, decision-making, and motor actions during human task performance are critical. Altogether, these qualitative facets of human task strategies should be part of the problem statement and must therefore be described precisely.

Once the problem domain has been identified, selected OO teams that consist of at least a business domain expert, an analyst, a facilitator, and a copy editor transcribe the problem statement as basis for OOD. An operative, functional, and logically correct problem statement requires time and effort because of possible iterations.

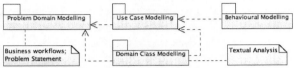

Figure 1 Problem domain modeling and its dependents

Benefits of the problem statement are its coexistence with workflows, description of the 'whole' versus isolation, reuse and consistency of artifacts, ability to derive test scenarios, and more. So, the more the indeterminate domain modeling, the more 'blurred' the whole, the higher the number of possible iterative cycles, the costlier and riskier processes and deliverables become (c.f., Figure 2).

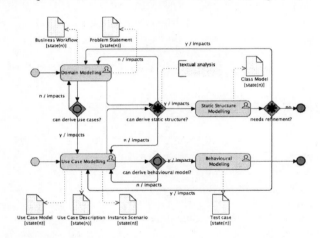

Figure 2 Impacts of the problem statement

Overall, our previous discussion underlines the importance of the degrees of precision and correctness of the descriptions containing the who's, what's, when's, how's, and implicit and explicit responsibilities because of the granularity that design and development require – notwithstanding project-related and organizational impacts.

5 A BASIC OVERVIEW OF METHODS

The awareness of the criticality of ambiguously written requirements and interdependent textual descriptions within the various OO models has brought forth numerous methods on formal, semi-formal, and informal heuristics planes. Finding the right balance between significance, criticality, and desired level of accuracy of the man-machine deliverable, business and software engineering methods, its purpose, readership and stakeholders, skills, time, and budget challenge the choice and need for the right method.

Work done in linguistic analysis and natural language processing is able to parse sentences that are written in a natural language. This area categorizes text input for requirements into *unrestricted descriptions*, *text specifications*, and *use case scenarios*:

Unrestricted descriptions do not restrict the written style of narration. Often, they are the result from interviews and are characterized by repetitions, redundant wording, synonyms, and fragmentations. Sentence syntax is arbitrarily complex and long in the range of approximately 100 sentences. Authorship and readership vary.

Text specifications are considered to be the most accurate descriptions in terms of natural language processing: Unnecessary wordings have to be removed, must be grammatically correct, and describe system requirements in logical and consecutive manners. Doing so requires text specifications to be comprehensible and exact based on a simplified syntax and unobstructed semantic references. A text specification for a small system contains approximately 50 sentences. Domain experts write text specifications for software specialists.

Use case descriptions and their scenarios are part of the most-simplified form of natural language. Their algorithmic and simple narrative styles need to list exact sequences of actions that are expressed in simple sentences. The volume of an average scenario should not exceed 10 – 15 sentences. Analysts write use case scenarios for design and development.

More formal methods are able to automate an entire OOA process because in a problem domain and in OO all models and views interrelate; the same is true for textual descriptions. Formal methods analyze linguistic information from informal specifications and abstract e.g., classes, objects, properties, actions, behaviors, actors, and data dictionaries. Some methods propose a mapping between the linguistic and conceptual worlds by having a set of predetermined structures from natural language to a set of grammar with defined syntax and semantics. Reasons are the potential unlimited ambiguities of natural language, vague assumptions, and unclear formulations. Other methods use e.g., semantic networks, ontologies, knowledge discovery, and mining techniques. However, for the average organization, these methods are usually too complex, complicated, and costly.

Nevertheless, the underlying rationale for the existence of the many methods is that all agree on those negative effects of ambiguous descriptions that lead to ill-specified systems, misconceptions between human goals, task performance, and system behaviors, the inability of engineers to design and develop systems from incomplete, imprecise, and isolated requirements and specifications.

6   CLASS, RESPONSIBILITIES, AND COLLABORATION

For a commercial organization that has realized deficiencies of ambiguous specifications and where failures of man-machine systems are not fatal, a heuristic-based approach will be more suitable than complex formal methods. Although the following relates to OO, principles that will be presented are equally applicable for a structured approach.

(Beck and Cunningham, 1989)'s informal heuristics-based method 'Class, Responsibilities, and Collaboration cards' (CRC) has been at the forefront for over two decades. CRC is a pragmatic method that anybody can apply without being a professional software engineer because it helps to identify the role of an object on three dimensions and can be applied during static structure modeling. Although the authors of the CRC method did at that time not address the problem statement, the importance of the context, naming conventions, and correctness of the language cannot be missed:

Class: "The class name of an object creates a vocabulary for discussing a design... object design has more in common with language design than with procedural program design... We urge... to find just the right set of words to describe our objects, a set that is internally consistent and evocative in the context of the larger design environment" (Beck and Cunningham, 1989, p. 2). Our extract of a problem statement of #2 of **Table 1** results in the candidate classes *'league'*, *'team'*, *'member'*, and *'contest'*.

Responsibility: "A responsibility serves as a handle for discussing potential solutions. The responsibilities of an object are expressed by a handful of short verb phrases, each containing an active verb. The more that can be expressed by these phrases, the more powerful and concise the design. Again, searching for just the right words is a valuable use of time while designing" (Beck and Cunningham, 1989, p. 2). In our example #2 of **Table 1** the 'team' *recruits* 'members'.

Collaborators: "We name as collaborators objects which will send or be sent messages in the course of satisfying responsibilities" (Beck and Cunningham, 1989, p. 2). Therefore, our 'team' of #2 of **Table 1** *collaborates* with the 'member'.

According to OOA, composing descriptions follows 4 basic steps such as:

1. To develop a summary paragraph that consists of Noun-Verb, Noun-Verb-Object, and Verb-Object variations.

2. To identify objects of interest by underlining all nouns in the summary paragraph.

3. To identify processes by circling all verbs in the summary paragraph.

4. To define attributes of objects by identifying adjectives or adjectival phrases and describing nouns which are objects in the solution space.

If an organization builds OO applications and yet uses conventional feature and functional requirements lists coupled with traceability matrices, applications may not behave they way they are expected to do. Unfortunately, hardly any organization uses methods to formulate e.g., features, requirements, problem statement, or use case scenarios. Therefore, they miss the strength of the pragmatic approach that heuristic-based methods like CRC have to offer. However, even the best method cannot "guarantee that the problem was defined correctly, i.e. the wrong requirements may have been stated properly" (Nunamaker et al., 1976, p. 685). Hence, apart from grammatical correctness, completeness, and more our enquiry towards less ambiguous requirements, descriptions should no longer exclude the essence of human task strategy and task performance because "[t]he formulation of a problem is often more essential than its solution, which may be merely a matter of mathematical or experimental skill" (Einstein and Infeld, 1938, p. 62).

7 CONCLUSIONS

Organizational praxis reveals that ambiguous informally written features, mutually dependent requirements, the absence and poor descriptions of workflows, processes, human task strategies, and the like foster gaps between expected behaviors by humans and system responses. These facets can lead to errors or even fatal failures, cause delays, affect human work performance, motivation, and productivity, increase costs, and more. Hence, an unambiguous, complete, semantically correct, and well-structured verbalization of a meaningful and situated problem, delta, feature, or requirement that must satisfy different levels of abstractions, degrees of granularity, and viewpoints towards a solution delivery is extremely challenging.

On one hand, this requires both the equal involvement of all stakeholders including decision-makers during feature and requirements elicitation, analysis, negotiation, decision-making, validation, design, and realization as well as the need to be able to communicate using one common language. On the other, this gives rise to questions as to the descriptive manner of deltas in view of quality of style, granularity, depth, the who's, what's, when's, and how's under the umbrella of a suitable methodology.

We have presented shortcomings of feature and requirements lists that are coupled with traceability matrices because the conventional functional engineering approach in use since the 1970's still dominates today's software engineering. Descriptions of such features or requirements are often no longer than one line, not understandable, decoupled islands from a larger 'whole', and completely ignore stimuli from goal-oriented human work activity e.g., human goals, task strategies, task performance, and human cognition and behavior while interacting with systems. We have therefore contrasted examples from the notion of the problem statement with conventional narrations of features and requirements.

For a better illustration, we have used the example of today's OO paradigm where the problem statement, which describes the problem or subject domain closely interrelates with descriptions of other models and views within the problem domain e.g., use case modeling, static modeling, and behavioral modeling. Contrary to conventional features and requirements on the basis of their structured approach, descriptions from OO perspectives are not decoupled islands, can be reused, are shareable, and are linked together within the universe of the problem domain.

Numerous methods have been developed to counterfeit ambiguous descriptions. In view of organizational suitability, we have 'revived' the CRC method and have postulated practical suggestions to better draft narratives. Although we have clearly presented advantages of the problem statement, its position, and relations within OO models, we do not expect that organizations will immediately change what they have been doing for years. However, we have shown practical suggestions that are equally applicable as of how to write better, less ambiguous, more complete, and grammatically more correct features and requirements.

Last but not least, we emphasize that human factors coupled with system identifiable interactions during goal-oriented human work activity should no longer be excluded from narrations. Our arguments are altogether not only the affair of engineering but also one of decision-makers, management, and business alike.

ACKNOWLEDGEMENTS

The authors would like to acknowledge the input and support received from Robert Gray, Thomas Wallmann, and Niels Bast.

REFERENCES

Beck, K. and W. Cunningham. 1989. A Laboratory For Teaching Object Oriented Thinking. Ed. N/A, *Object-Oriented Programming, Systems, Languages, and Applications (OOPSLA 1989)*, New Orleans, Louisiana 1-6.

Bedny, G. and W. Karwowski 2007. *A Systemic-Structural Theory of Activity: Applications to Human Performance and Work Design*. Taylor & Francis, Boca Raton.

Bedny, G.Z., W. Karwowski and O.-J. Jeng. 2003. Concept of Orienting Activity and Situation Awareness. Ed. N/A, *Ergonomics in the Digital Age, IEA Congress 2003*, Seoul **6:** 447-450.

Bedny, I.S., W. Karwowski and G.Z. Bedny. 2010. A Method of Human Reliability Assessment Based on Systemic-Structural Activity Theory. *International Journal of Human-Computer Interaction*, **26:** 377-402.

DeMarco, T. 1996. Requirements Engineering: Why Aren't We Better at It? *2nd International Conference on Requirements Engineering (ICRE 1996)*, Colorado Springs 2-3.

Einstein, A. and L. Infeld 1938. *The Evolution of Physics*. Simon & Schuster, New York.

Nunamaker, J.F., Jr., et al. 1976. Computer-aided analysis and design of information systems. *Communications of the ACM*, **19:** 674-687.

Taylor, R.N. and A. van der Hoek. 2007. Software Design and Architecture - The once and future focus of software engineering. Ed. N/A, *2007 Future of Software Engineering*, Washington, D. C.

von Brevern, H. and K. Synytsya. 2006. A Systemic Activity based Approach for Holistic Learning & Training Systems. *Special Issue on Next Generation e-Learning Systems: Intelligent Applications and Smart Design. Educational Technology & Society*, **9:** 100-111.

von Brevern, H. and K. Synytsya. In press. Towards 2020 with SSAT. In *Science, Technology, Higher Education and Society in the Conceptual Age*. Eds. Marek, T., W. Karwowski and J. Kantola. Taylor & Francis, London, New York: pp. tbd.

CHAPTER 5

Overloading Disks onto a Mind: Quantity Effects in the Attribution of Mental States to Technological Systems

O. Parlangeli, S. Guidi, R. Fiore Farina
Communication Sciences Department
University of Siena
Siena, Italy
oroparla@gmail.com

ABSTRACT

The study reported here has been structured to assess whether the number of very simple elements - little disks displayed on a screen - may affect the process of attributing mental states to these elements. One hundred and twenty subjects, allotted in three different groups, participated in an experiment in which they had to interact either with two, six, or eighteen disks in a videogame. Subjects had simply to click on the disks in order to gain scores.

Results show that the number of disks displayed affects the attribution of some mental states. Specifically, beliefs about the disk having "attention", "awareness" and "memories" were lowest in the condition with an intermediate (six) number of disks. This seems to suggest that different processes may be at play mediating the way in which mental state attributions are affected by the number of elements in the system.

Keywords: mental states attribution, cognitive overload, theory of mind

1. INTRODUCTION

Research in cognitive ergonomics has usually assumed that human-computer interactions are based on the reciprocal elaboration of some interpretative models. Expectations about the way a system works come from mental models progressively elaborated by the user through consideration of the computer's behavior. On the other side, knowledge about the user is inserted inside the user's model, which informs the designers' activity (Carroll and Olson, 1987; Allen, 1997). These models are traditionally considered as models of processes, sometimes they are incomplete or not coincident with the actual functioning of the system, but quite often, are sufficiently adequate to give course to a productive interaction. More specifically, the user is believed to elaborate mental models that would be used as a means to make predictions, to produce explanations, and to provide diagnosis about the behavior of the system (Allen, 1997).

Being mental models are essentially cognitive tools for the interpretation of complex system's behavior, they should be elaborated based upon more or less reasonable causal laws, these interpretive rules being directly referred either to scientific or to naïve physics (Bozzi, 1991; Molina et al., 1994). Contrary to this purely mechanical perspective, however, we all have been witnesses of interactions between human beings and technology in which informative systems are nearly treated not as mere machines but more as persons. This implies that for the user, in some circumstances, the behavior of the system could be explained by other reasons than physical and electronic determinants. These kinds of behaviors – people speaking, smiling, caressing, quarrelling with their computers – have been generally considered as an oddity, though they are very frequent and relatable to many different systems.

In the past it has been shown that people tend to consider as human agents those systems that show some changes, and, since the analyses of Premack and Woodruff (1978), it has been claimed that this phenomenon involves a fundamental bias, that is the bias that brings us to attribute mental states to other people, a very complex process that has been described as the elaboration of a Theory of Mind. A very relevant theory about this phenomenon has been formulated by Dennett (1987), which maintains the existence of three different levels of interpretations that are adopted by human beings while they interact with everything in the world. At a first level we have the possibility to explain the behavior of any system referring to more or less validated scientific laws (physical stance). At a higher level, the design stance may be assumed, thus accounting for the designer perspective in order to find the reasons of the behavior of a system. The last level is the one in which systems are gifted with a mind, in which intentions, desires and belief are assumed to explain why a system behaves just the way it does (intentional stance).

When interacting with a technological system, a purely rational perspective should bring human beings to adopt the second level of abstraction, the design stance. However, research has highlighted notable exceptions to this ideal behavior. Some studies have shown that people have a tendency to anthropomorphization (Baron-Cohen, 1995a) when systems show characteristics such as self-propulsion

(Baron-Cohen, 1995b), movement along a trajectory or apparently directed toward a target (Dittrich and Lea, 1994), and also when they show movements that are similar to those of human beings (Morewedge et al., 2007).

In addition, some recent studies (Steinbes and Koelsch, 2009) in the field of neuropsychology have shown that when people believe that they are interacting with an artifact, that is with a human product, it is possible to record a cortical activity that is in the same cortical network (anterior medial frontal cortex – superior temporal sulcus – and temporal poles) that is usually activated during processes of mental states attribution. And other studies have shown that the bias to attribute mental states is enhanced when anthropocentric knowledge may be applied to the systems, when there is a high motivation to understand the behavior of other agents and when, for many different reasons, there may be the desire for social contacts (Epley et al., 2007).

To sum up, the attribution of mental states to technological systems does not appear to be a sporadic and strange phenomenon. On the contrary, it seems quite common and rooted in the same cognitive and neurological processes that are at the basis of human-human social interactions.

Assuming this perspective, some studies on the elaboration of a theory of mind in relation to the behavior of some robots (Wang et al., 2006; Terada et al., 2007) showed that human beings are more prone to attribute mental states to robots that are reactive to user actions, and whose possible uses may be more easily detected.

Parlangeli et al. (2012) have very recently underlined how the natural bias that brings human beings to elaborate a theory of mind may come into play even in relation to very simple artificial elements. In the study subjects had to interact with disks that were just blinking on a computer screen. Trough a post-test questionnaire it was shown that participants in the experiment considered those simple disks has having, at some level, intentions, strategies, consciousness, and a mind. In that study differences in the attribution of each mental state were found, and some contextual factors, as the number of disks displayed, appeared to have a possible role on the process of mental state attribution.

The study reported here has a twofold aim. On the one side it has been structured to better assess whether the number of element displayed on a screen may affect the process of mental state attribution to those elements. On the other side, the attribution of other mental states such as attention and memories will be explored for the first time.

2. THE STUDY

2.1 Methods

2.1.1 Participants

120 participants took part in the experiment. 57.5% of the subjects were male, and 42.5 % female. The age of participants was between 15 and 30 years old, but

the majority (64.2 %) had between 20 and 25 years, while 30.8% had between 15 and 19 years, and only about 5% was more than 25 years old. Most of participants (92.5%) were either high school or undergraduate students.

All the subjects voluntarily participated in the experiment, and were unaware of the real aim of the study.

2.1.2 Procedure and materials

During the experiment participants were asked to play a video game in which they had to click on some colored disks in order to gain score. The position of the disks was randomly determined at the onset of the game, and randomly changed every 1 or 2 seconds. The overall duration of the game was 80 seconds. Half the disks were red and measured 16 mm in diameter. The other half was instead blue and had a diameter of 20 mm. The position of the red disks was varied every 1 second, and the one of the blue disk every 2, and participants were told that they would have gained higher scores by clicking on the red than on the blue disks.

Participants were randomly assigned to one of three different conditions, so that 40 subjects served in each condition. In the first condition there were only 2 disks on the screen; in the second condition the disks were 6 and in the third they were 18.

Before playing the game, participants were asked to fill in a survey comprising three sections. The first section included some demographical questions about participants' age, gender, education and employment status. The second part contained an Italian translation of the TIPI inventory (Gosling et al., 2003), in order to assess their scores on five personality traits: agreeableness, openness to experience, extraversion, conscientiousness and emotional stability. The last section of the survey was the Italian version of the Media Multitasking Index (Ophir et al., 2009), a questionnaire developed to assess the mean number of media a person simultaneously uses when consuming other media.

After the game, participants were asked to fill in a second survey, in which they had to report, on a seven-point scale, if they had thought the disks "had their own strategy", "were aware of what was happening", "were paying attention to what was happening", "had their own intentions", "had memories of what had happened", and "had their own mind". In addition, they were asked to express their degree of agreement (again on a seven-point scale) with the some control statements: "the movements of the disks were random", "the game was regular", and "some disks had more intentions than others".

2.2 Results and discussion

We first performed some descriptive statistics on the collected data, and excluded from further consideration those coming from subjects that reported maximum disbelief (i.e. 1) about *all* the questions concerning the system having any kind of mental state. The reason for this choice is that we were interested in assessing how many participants believed, even to a minimum degree of confidence, that a videogame could have one or more kinds of mental states. We thus counted them and removed those completely "skeptical" subjects who were not prone to attribute any mental state or property to a game. Following this criterion, anyway, only 7,5% of data were excluded.

In figure 1, are reported the mean belief scores for the different mental states, averaged across experimental conditions. As it can be seen for most of the mental states considered average scores are quite low, although always higher than the bottom end of the scale, representing total disbelief, as confirmed also by t-tests (p<.0001). However, it is also evident that different kinds of mental states were judged to differently belong to the disks. These were believed to have a strategy more than to have awareness or intentions, and the lowest ratings were relative to the disks having a mind.

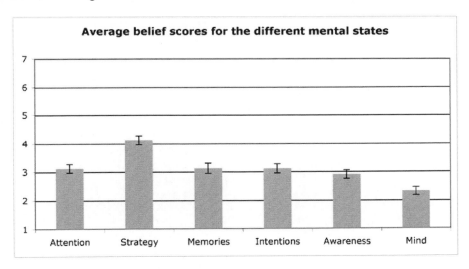

Figure 1. Average belief score of participants relative to the different mental state considered (across conditions).

We performed a multivariate between subjects analysis of variance (MANOVA) on the response data, using the number of disks as the independent variable, and each of the mental states as dependent variable. The overall multivariate test was significant (p<.01), but the analysis of the univariate between-subjects effects revealed significant effects of condition (number of disks) only on three dependent variables: the belief scores for "attention" ($F(2,108)=4,016$: $p<.05$), "memories" ($F(2,108)=5,533$: $p<.01$) and "awareness" ($F(2,108)=4,059$: $p<.05$). We also tested whether there were difference between male and females in the belief scores, but no significant effects of gender were found on any of the dependent variables. We also conducted another MANOVA on the belief scores for the control questions, and found that the belief scores about the disks having "random movements" were significantly affected by the number of disks ($F(2,108)=5,189$; $p<.01$).

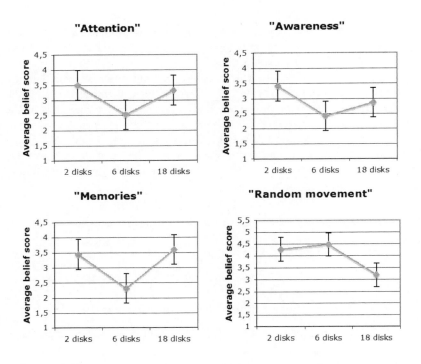

Figure 3. Average belief scores of the subject for three mental states and a control variable as function of the number of disks present in the game.

In figure 3 are reported the average belief scores for these mental states and for the control variable ("random movements") as function of the number of disks. As it can be seen, the intermediate 6-disk condition was the one in which belief scores were lowest for each mental states. Post-doc pairwise comparisons (Bonferroni corrected) showed that belief scores about "memories" in the 6-disk condition were lower than in both the other conditions ($p<.05$ vs 2-disk; $p<.01$ vs 18-disk). For "awareness", instead, belief scores in the 6-disk condition were only significantly lower than in the 2-disk condition, while for "attention" belief scores in the 6-disk condition were significantly lower in than in the 2-disk condition and marginally also lower than in the 18-disk one. As for the belief scores about the "random movements" of the disks, pairwise comparisons revealed lower scores in the 18-disk condition than in the other two experimental conditions.

We analyzed the correlations between the belief scores for the different mental states. The correlation matrix is reported in table 1. As it can be seen in the matrix, all the belief scores about the different mental states were significantly correlated with each other, with Pearson coefficients ranging from .2 to .68. The highest correlation was found between "awareness" and "attention" ($r=.68$; $p<.0001$). Belief scores about "attention" and "memories" were also moderately correlated ($r=.54$;

p<.0001), and so were scores about "memories" and "awareness" (r=.49; p<.0001). Moderate significant correlation coefficients were also found between beliefs about "strategy", "intentions" and "mind". When we considered the correlations of the control variables with the belief scores about the mental states, we found only negative significant correlations between the belief scores about the disk having "random movements" and those relative to "memories" (r=-.26; p<.01), "strategy" (r=-.29; p<.01), and "intentions" (r=-.2; p<.05).

Table 1. Bivariate correlations between the belief scores for the different mental states. (All the correlations are significant either at .05 or at .01 level)

	Awareness	Memories	Attention	Strategy	Intentions	Mind
Awareness	–	0.488	0.681	0.199	0.296	0.263
Memories		–	0.536	0.429	0.349	0.273
Attention			–	0.270	0.273	0.292
Strategy				–	0.445	0.314
Intentions					–	0.453
Mind						–

We also conducted a factor analysis on the belief scores for the different mental states, using principal components and oblimin rotation, to investigate the structure of the participants' attributions about mental states to the system. Not surprisingly, at the light of the correlation matrix above, the analysis yielded a two-factor solution (explaining about 67% of the variance), which clearly segregated (table 2) beliefs about "awareness", "attention", and "memories" from those about the other three mental states. We computed participants' factor scores, and repeated the MANOVA testing for differences between the experimental conditions. Consistently with the previous analysis, the effect of the number of disks was significant only on the factor relative to "awareness", "memory" and "attention", whose scores were significantly lower in the 6-disk than in the other two conditions.

The analysis of the correlations of the belief scores with the five personality traits assessed with the TIPI inventory only found a mild significant correlation between belief scores about "memories" and participants' "conscientiousness" (r=.27; p<.01). When we looked at the correlations of the factor scores with the personality traits, however, we also found a significant mild correlation between the "attention-memory-awareness" factor scores and participants' scores of "agreeableness" (r=.19; p<.05).

Finally, we looked at the correlations between the belief and factor scores and the average number of media used in multitasking (MMI). Only belief scores about "intentions", however, was significantly correlated with MMI scores, and the size of the correlation was quite small (r=.19; p<.05).

Table 2. PCA estimates of the (oblique) factor loadings for the belief scores of the different mental states

Mental states	Factor loadings	
	Factor 1	Factor 2
Awareness	.917	
Attention	.910	
Memories	.639	
Intentions		.824
Strategy		.765
Mind		.724
Eigenvalue	2.872	1.114
Variance explained (%)	47.9%	18.6%

3. CONCLUSIONS

The results of the present study confirm previous findings by Parlangeli et al. (2012) that attribution of mental states doesn't seem an all or nothing process. Participants did not simply believe that the disks either had or did not have a mind. Conversely, the majority of our subjects showed graded beliefs about the mental states of the system. Overall, they expressed doubts, more or less strong, suspects, more or less evident, about the possibility that the elements of the system they interacted with had the mental states considered in the study.

Consistent with previous research, also in this study it was found that participants believed with different degrees of confidence in the presence of different mental states in the system. Talking about attention, awareness or strategy can have different meaning even with reference to very simple elements of a technological system, such as the disks used in the videogame of our experiment.

For the first time, however, our results show an interesting distinction between different kinds of mental states. In our subjects' responses, in fact, there seems to be a clear distinction between beliefs pertaining to what we could call "cognitive systems," such as memory, attention or awareness, and the more "procedural" mental states relative to "cognitive functions," such as the ability to follow a strategy, to pursue intentions and to express mental abilities.

It is precisely in relation to the first kind of mental states, those having a more systemic nature, that attribution seems to be affected by the quantity of the elements subjects are interacting with. In the condition with six disks, in fact, belief scores about these mental states were lowest, reflecting higher degrees of doubt. And though in light of what is currently known about the process of mental states attribution it is difficult to fully explain this result, our study seems to suggest that different factors might be involved. On the one hand, in fact, it might be that a low number of disks (such as in the first condition in which there were only two of them) could make it easier for subjects to consider each disk individually. On the other hand, instead, the higher number of elements displayed in the third condition (eighteen) could require more attention from our subjects, which in turn could bring them to assume a default mentalistic disposition. These hypotheses, anyway, are clearly yet speculative, and will thus require further empirical investigation.

REFERENCES

Allen R.B. (1997) Mental models and user models. In M.G. Helander, K. Landauer and P.V. Prabhu (eds.) Handbook of Human Computer Interaction. North Holland, Amsterdam, pp. 49-64.

Baron-Cohen, S. (1995a) Mindblindness: An Essay on Autism and Theory of Mind, The MIT Press, Cambridge, Mass.

Baron-Cohen, S. (1995b) The eye direction detector (edd) and the shared attention mechanism (sam): Two cases for evolutionary psychology," in Joint Attention: Its Origins and Role in Development, C. Moore and P.J. Dunham (Eds.), Lawrence Erlbaum Associates, 3, pp. 41–59.

Bozzi, P. (1991), Fisica ingenua, Garzanti, Milan.

Carroll, J.M., and Olson J.R. (1987), Mental models in human-computer interaction, National Academy Press, Washington, DC.

Dennett D.C. (1987) The Intentional Stance, Bradford Books/MIT Press, Cambridge, Mass.

Dittrich, W.H. and Lea, S.E.G (1994) Visual perception of intentional motion, Perception, 23, 3, pp. 253–268.

Epley, N., Waytz, A. and Cacioppo, J.T. (2007) On seeing human: A three-factor theory of anthropomorfism. Psychological review, 114, 4, 864-886.

Gosling, S.D., Rentfrow, P.J. and Swann, R.B. Jr. (2003) Journal of Research in Personality, 37, 504-528.

Molina, M., Van de Walle, G.A., Condry, K., and Spelke E.S. (2004) The animate–inanimate distinction in infancy: Developing sensitivity to constraints on human actions, Journal of Cognition and Development, 5, pp. 399–426.

Morewedge, C.K., Preston J. and Wegner D.M. (2007) Timescale bias in the attribution of mind. Journal of personality and social psychology, 93, 1, 1-11.

Ophir E., Nass C.I., Wagner A.D. (2009) Cognitive control in media multitaskers. Proceedings of the National Academy of Science USA, 106, pp. 15583–15587

Parlangeli, O., Chiantini, T. and Guidi, S. (2012) A mind in a disk: The attribution of mental state to technological. Work, 41, pp. 1118-1123.

Premack, D. and Woodruff, G. (1978) Does the chimpanzee have a theory of mind? The Behavioral and Brain Sciences, 4, pp. 515–526.

Steinbeis, N., and Koelsch S. (2009) Understanding the intentions behind man-made products elicits neural activity in Areas dedicated to Mental State Attribution. Cerebral Cortex, 19, 3, pp. 619-623.

Terada, K., Shamoto, Y., Mei, H., and Ito, A., (2007) Reactive Movements of non-humanoid robots cause intention attribution in humans. Proceedings of the 2007 IEEE/RSJ International Conference on Intelligent Robots and Systems San Diego, CA, USA, Oct 29 - Nov 2, 2007.

Wang, E., Lignos, C., Vatsal, A., and Scassellati, B., (2006) Effects of head movement on perceptions of humanoid robot behavior, Proceeding of the 1st ACM SIGCHI/SIGART conference on Human-robot interaction, (2006), pp. 180–185.

CHAPTER 6

Effectiveness of Two-Channel Bone Conduction Audio Communication System in Speech Recognition Tasks

Maranda McBride
North Carolina A&T State University
Greensboro, NC, USA
mcbride@ncat.edu

Rachel Weatherless and Tomasz Letowski
U.S. Army Research Laboratory
Aberdeen, MD, USA

ABSTRACT

The objective of this study was to assess the effectiveness of a bone conduction (BC) communication interface in two-channel speech recognition tasks. Twenty-four listeners participated in a speech-on-speech masking task to assess the effectiveness of a two-channel BC listening device designed for speech perception in a two-channel communication system. The task involved listening to either one target sentence presented randomly to each ear (monaural condition) or two sentences (one target, one competing) presented simultaneously to separate ears (dichotic condition) via a BC headset or air conduction (AC) headphone set. Each participant listened to two trial sets during the study. In each trial set either twenty individual or twenty pairs of sentences were presented to listeners who were asked to respond by selecting the appropriate words from word lists presented on the computer screen. The speech signals were sentences from the Synchronized Sentence Set (S^3) test which enables simultaneous presentation of one target (T) sentence and up to three competing (C) sentences during a test condition. Results of the study indicate that listeners using the BC headset perform significantly better in the monaural condition than in the dichotic condition; however, from an operational standpoint, listeners still had acceptable performance under both conditions. A comparison of the AC and BC systems under the same dichotic conditions indicated no significant differences in performance of the two types of systems. This implies that the BC system is as effective as the AC system in two-channel speech-on-speech masking tasks.

Keywords: bone conduction, multichannel communication, competing speech

1 INTRODUCTION

The use of bone conduction (BC) listening devices has become more common in the last 10 years in both military and commercial settings. These devices transmit sounds by bypassing the outer and most of the middle ear structures to stimulate the components of inner ear by sending audio frequency vibrations to the bones in the skull. One of the primary benefits of BC devices is that they do not disrupt the listener's situational awareness as much as the standard air conduction (AC) devices such as headphones. This is because the BC devices do not impede the perception of interaural time and intensity differences in auditory signals since they leave the listener's ears open (McBride, Letowski, & Tran, 2008). This is an important property for communication systems used in complex and potentially hazardous surroundings, such as battlefield environments, where uncompromised situation awareness is critical for soldier safety and effectiveness.

In military operations, BC devices offer several advantages over traditional AC devices. For example, because they are small and light in weight, incorporation into military headgear is relatively effortless (McBride, Hodges, & French, 2008). However, the effectiveness of such devices under certain operational conditions has come under scrutiny. For instance, previous research studies have shown that listening task performance with these devices is dependent upon where the vibrators are located on the head (McBride, Hodges, & French, 2008; McBride, Letowski, & Tran, 2008). In other words, some skull locations transmit BC signals better than others.

Another concern regarding BC headsets is their ability to overcome the low intracranial attenuation of bone conducted signals and be used in multi-channel communication systems. Due to a common perception among military personnel that BC is not a very conducive means for multichannel communication, two-channel BC system have yet to be considered for military applications even though some studies have indicated that 3D audio virtual displays can be used with the BC systems (e.g., MacDonald et al., 2006). The purpose of the current study was to determine if two-channel BC systems can be used as effectively for two-channel communication system as two-channel earphone-based systems and, if not, what the specific disadvantages of this system are.

2 METHOD

Participants

A total of twenty-four participants (12 male and 12 female) were recruited for this study. Their ages ranged from 21 to 59 years. Participants were volunteers from the Aberdeen Proving Ground, MD community. All participants were required to undergo a hearing test, which was performed using a Madsen Orbiter 220 audiometer and TDH-50 headphones. The hearing tests took place in a sound-treated booth with noise levels that comply with ANSI S3.1 (1999) standards for earphone listening. Normal hearing was based on pure-tone AC hearing thresholds of 20 dB HL or lower for audiometric frequencies from 250 to 8000 Hz, inclusively, (ANSI S3.6, 2010) and no history of otologic pathology. In addition, to ensure symmetric hearing, the difference between the pure-tone hearing levels for both ears at any test frequency could be no greater than 10 dB.

Test Material

After passing the hearing test, participants were seated at a listening station in a large audiometric room meeting ANSI S3.1-1966 criteria for open ear listening. The listening station consisted of a Dell laptop computer, an Oiido SA2 BC two-channel headset, and a set of Telephonics TDH-39 headphones. The selection of the two specific transducers was based on their relative high quality for speech communication reported in the previous studies. To compare the effectiveness of both two-channel communication systems, the Synchronized Sentence Set (S^3) test (Abouchacra et al., 2001; Abouchacra et al., 2009) was used. The Synchronized Sentence Set (S^3) has been known previously as the SATASK database (e.g., Abouchacra et al., 1997; Stern et al., 2006) but the name has been change due to the name conflict.

The S^3 is a speech recognition test designed by the Army Research Laboratory to measure speech perception performance in speech-on-speech masking conditions. The speech signals used in the S^3 test were from a set of 2592 sentences (10 syllables each) constructed from 30 token words (see Table 1). Each token word was recorded by four different male talkers. When presented together, sentences are temporally synchronized in such a way that the key words in the individual sentences begin and end simultaneously. Each sentence consists of words spoken by only one of the four talkers. The features of the S^3 software allow the researcher to simultaneously present one target (T) sentence and up to three competing (C) sentences during a single trial. The experimenter can independently route the T- and C-sentences through one or more transducers (e.g., bone conduction vibrators) to create a variety of virtual 3D audio divided and selective attention tasks. In the current study, only two channels (one used to present the T-sentence and the other used to present the C-sentence) were used.

Table 1: S3 Sentences Words

NAME	NUMBER	COLOR	OBJECT
Mike	1	Black	Ball
Nate	2	Blue	Cup
Ron	3	Brown	Fork
Troy	4	Gray	Key
	5	Green	Kite
	6	Pink	Spoon
	8	Red	Square
	9	White	Stair
			Star

Procedure

The participant's task was to listen to either one S^3 sentence presented to one ear or two S^3 sentences presented simultaneously but separately to each ear. The listener was asked to respond only to the sentences starting with the name "Troy." All sentences had the following structure: "(NAME) mark the (NUMBER) on the (COLOR) (OBJECT)". The T-sentence would be presented through only one transducer while the other transducer would present a C-sentence starting with a different name. Both sentences contained four token words (a name, number, color, and object) and the participant was instructed to identify the token words only for the T-sentence, regardless of the perceived location of the sound source or the voice of the talker. If only one sentence was presented, it was always a T-sentence.

Each participant took part in one of two experiments each consisting of two test conditions. In both experiments, participants were asked to respond to each sentence beginning with the name "Troy" by using a mouse to select the three token words they heard from dropdown lists provided through the computer interface (Figure 1). After completing the sentence, the participant clicked on the Next Sound button to present the next signal (or set of signals) until all 20 presentations were completed.

The first experiment (Experiment 1) required participants to listen to either one target sentence presented randomly to each ear (i.e., monotic condition) or two sentences (one T-sentence, one C-sentence) presented simultaneously to separate ears (i.e., dichotic condition) via a BC headset. Half of the participants were exposed to the dichotic condition first while the other half listened to the monotic condition first. For each condition, either 20 single sentences (monotic condition) or 20 pairs of sentences (dichotic condition) were presented.

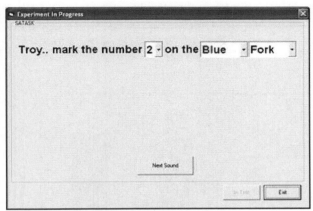

Figure 1: Response Screen

The second experiment (Experiment 2) exposed participants to the dichotic condition only with both a BC headset and headphones. Half of the participants were tested with the BC headset first while the other half used headphones first.

A single trial set consisting of one listening condition lasted approximately five minutes. The words in each sentence were randomly selected from the S^3 corpus according to the S^3 rules. In all conditions, the T-sentence was presented 10 times to the left and 10 times to the right ear in a random order to avoid data contamination by ear advantage effect (Repp, 1977; Divenyi & Efron, 1979).

3 DATA ANALYSIS AND RESULTS

A graphical depiction of the data obtained in both experiments and averaged across all participating listeners is provided in Figure 2. The data were analyzed using two repeated measure one-way ANOVAs both with two levels of the independent variable (Condition). For Experiment 1, the monotic (BC Mono) and dichotic (BC Dich) speech recognition scores were compared for each participant. The analysis for Experiment 2 compared the AC headphone (AC Dich) and BC headset (BC Dich) speech recognition scores for each participant. For each trial, the speech recognition score was the percentage of complete sentences identified correctly. Prior to performing the analyses, the speech recognition scores were converted to rau units in order to reduce the ceiling effect often associated with using a percentage scale (Studebaker, 1985).

The results of the data analysis for Experiment 1 indicated a significant difference between the monotic condition and the dichotic condition ($F_{1,11} = 17.04$, p-value < 0.01). Based on the data, the speech recognition scores for the monotic condition (mean = 99.58%, standard deviation = 1.44%) was significantly higher than the scores for the dichotic condition (mean = 87.92%, standard deviation = 12.33%).

The results of Experiment 2 indicate no significant difference between the dichotic condition scores for the BC headset and the headphones ($F_{1,11}$= 0.53, p = 0.48). The mean speech recognition score for the BC headset was 85.83% (standard deviation = 14.28%) while the mean score for the headphone set was 83.33% (standard deviation = 10.73%).

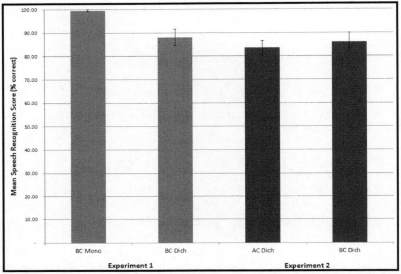

Figure 2: Mean speech recognition score per experimental condition

4 DISCUSSION AND CONCLUSIONS

The purpose of this study was to assess the effectiveness of a BC headset in a two-channel speech communication system. The tests were concerned with the masking effect of speech presented through one channel on speech heard through another. While such crosstalk/masking has been studied previously for headphone communication it has not been investigated for BC communication systems. It is important to note that the amount of the observed crosstalk/masking is dependent on both the temporal and spectral properties of both speech signals and the informational similarity of the messages. Therefore, the quantitative comparison between specific two-channel communication systems has to be limited to the same speech material. This greatly limits the inferences that can be drawn from the previous studies in which different speech material and different communication systems were used.

The main finding of the study is that the listeners' performance was very similar regardless of whether the listeners used a two-channel AC interface or two-channel BC interface. Therefore, it can be inferred from these data that the BC system performs as well as the AC system in the dichotic two-channel selective listening tasks. In both cases the listeners demonstrated acceptable performance per MIL-

STD-1472G (Department of Defense, 2012), which states that the minimal acceptable speech intelligibility (recognition) score for military communication is 75%. This criterion applies to speech-in-noise and multichannel communication. It is also important to note that the BC selective attention scores obtained in both experiments were almost identical (~88% and ~86%, respectively) indicating robustness and repeatability of the data obtained.

The results of the study further show that the listeners were able to recognize about 99.6% of the speech items in a single-channel speech recognition task using the BC interface. This score compares well to consistent 100% scores obtained informally under similar conditions for TDH-39 earphones by a number of our listeners. Both of these scores are much higher than the 91% criterion recommended for critical military communications (e.g., single channel communication in quiet) as specified in MIL-STD-1472G (Department of Defense, 2012).

Based on the data obtained and within the constraints of the conducted study, it can be concluded that the BC system used in this study was as efficient for military communications in both single-channel and two-channel applications as the headphone system used for comparison. The fact that the headphone system used in this study provided perfect scores consistently for single-channel listening may serve as indication that the BC system evaluated has sufficient quality to be used in lieu of any headphone system. However, more studies need to be conducted using a wider range of BC systems and the joint use of BC systems and hearing protection systems in noisy listening conditions to further support this statement.

REFERENCES

Abouchacra, K. M., Tran, T. V., Besing, J. M., and Koehnke, J. D. (1997). Performance on a selective listening task as a function of stimuli presentation mode. *Association for Research in Otolaryngology Abstracts*, 53.

Abouchacra, K., Breitenbach, J., Mermagen, T., & Letowski, T. (2001). Binaural helmet: Improving speech recognition in noise with spatialized sound. *Human Factors*, *43*(4), 584-594.

Abouchacra, K., Letowski, T., Koehnke, J., and Besing, J. (2009). Clinical application of the Synchronized Sentence Set (S^3). *Proceedings of the 16th International Congress on Sound and Vibrations (on CD)*, pp. 1-8. Cracow (Poland): IIAV. July 5-9.American National Standards Institute (1999). Maximum Permissible Ambient Noise Levels for Audiometric Test Rooms. ANSI S3.1-1999(R2008). New York (NY): ANSI.

American National Standards Institute (2010). Specifications for audiometers. ANSI S3.6-2010. New York (NY): ANSI.

Brungart, D. S., & Simpson, B. D. (2002). The effects of spatial separation in distance on the informational and energetic masking of a nearby speech signal. *The Journal of the Acoustical Society of America, 112*(2), 664-676.

Department of Defense (2012). Human Engineering Design Criteria for Military Systems, Equipment and Systems. MIL-STD-1472G.. Washington (DC): DOD

Divenyi, P.L. & Efron, R.E. (1979). Spectral versus temporal features in dichotic listening. *Brain and Language*, 7(3), 375-486.

MacDonald, J., Henry, P., & Letowski T. (2006) . Spatial audio through a bone conduction interface. *International Journal of Audiology*, 45: 595-599.McBride, M., Hodges, M., & French, J. (2008). Speech intelligibility differences of male and female vocal signals transmitted through bone conduction in background noise: Implications for voice communication headset design. *International Journal of Industrial Ergonomics, 38*(11-12), 1038-1044.

McBride, M., Letowski, T., & Tran, P. (2008). Bone conduction reception: Head sensitivity mapping. *Ergonomics, 51*(5), 702-718.

Repp, B.H. (1977). Dichotic competition of speech sounds: The role of acoustic stimulus structure. *Journal of Experimental Psychology: Human Perception and Performance*, 3 (1), 47-50.

Stern, R.M., Trahiotis, C., & Ripepi, A.M. (2006). Fluctuations in amplitude and frequency enable interaural delays to foster the identification of speech-like stimuli. In: P.Divenyi, S. Greenberg, & G. Meyer (eds.), *Dynamics of Speech Production and Perception*, pp. 143-151. Amsterdam (Netherlands): IOS Press.

Studebaker, G.A. (1985). A" rationalized" arcsine transform. *Journal of Speech and Hearing Research, 28*(3), 455-462.

CHAPTER 7

Cognitive Ergonomics: The User and Brand in an Informational Relationship

Sabrina Talita Oliveira, Adriano Heemann and Maria Lucia Okimoto

Federal University of Parana
Curitiba, Brazil
binah.oliveira@gmail.com

ABSTRACT

The relationship between user and brand is still an area little explored by cognitive ergonomics. This motivates an improved analysis and discussion about the process of collaborative design in the relationship between user and brand during the post-product development. Thus, this paper presents an analysis of communication between user and brand and multisensory effects resulting from this process that can stimulate and induce the purchase. This work is based on an exploratory literature concerning the Product Development Process (2) and promotes theoretical review of collaborative design (3), in order to identify the strategies adopted in the management of sensory communication of a brand, the process of interaction and sharing in the relationship between user and brand in the building the product image (4). Correlates studies and presents qualitative results (5) that show a new approach to the cognitive aspects of multisensory brand. This approach can help designers in the management of communication to enhance the communication performance during insertion of the product in commerce.

Keywords: relationship, user, brand

1 INTRODUCTION

This paper focused on the Unified Model of Reference for the Product Development Process - PDP (Rozenfeld et al., 2006 – our translation) that includes the final stage so called post development macro-phase. This development model was chosen because the goal of this paper is to present the process of collaboration and interaction in the user experience with the product.

According to Rozenfeld et al. (2006) the reference model for PDP is divided into three macro-phases: predevelopment, development, and post-development, each of which is subdivided into stages (or phases); and these, in turn, into activities. This study approaches the least macro-phase of development process, namely, post development.

This paper discusses the process of collaborative design in the post development of products. In the post development is the influence provided by the cognitive universe of the brand from the relationship between user and product and the building brand image. Thus, this approach discusses the mutual relationship of communication that appears in the first instance from the intangible values expressed by the brand and later, the interpretation of these values by the user.

It argues that the purchase of products transcends the physiological needs of the individual. In this perspective, sensory and cognitive strategies applied in the atmosphere of the brand can be an aggravating factor for the ideal symbolic articulation of the user.

Within this context and from the study of the Unified Model of Reference of PDP, the paper emphasizes collaboration in post-product development and sensory strategies of the brand, from communication management. That is, presents a new approach to collaborative design issue because it instigates collaboration also occurs in the post-product development, not only during development projectual.

This approach offers the vision of the consumer interacting with the sensory aspects of the brand atmosphere in order to confirm the construction of the image of the product. In the practical field, this analysis can stimulate collaboration in design during the post development, and adoption of cognitive strategies in the management of sensory communication. These strategies can determine the level of interaction, sharing and collaboration in the relationship between user and brand.

1.2 Methods

This paper is based on an exploratory literature on development methods projectual in design. Sets the Unified Model of PDP (Rozenfeld et al, 2006) as a method projectual basis for this study, because this method includes the final stage of post development, unusual in classical projective methodologies.

Later, promotes theoretical review of the design concepts related to collaborative cooperation, sharing and interaction of individuals, information and tasks between professionals. It favors the adoption of a new reasoning, because it shows that the collaborative process can also occur in the post development.

Argues about sensory branding and brand management. Correlates these studies and presents qualitative results that point to a new approach. Thus, demonstrates cognitive aspects that can assist designers in brand management, in order to optimize the communication performance of the product.

Finally, presents a discussion on collaboration in the post development, where the user interacts with the atmosphere of the brand.

2 DESIGN AND METHODOLOGY PROJECTUAL

The design acts as a systematic creative process, with various control variables, is to develop products that meet the needs of use "practical, aesthetic and symbolic" (Lobach, 2001).

In this sense, the project design needs to be done from projectual methodology (Villas-Boas, 1998). The method must include the stage of the genesis of the product, must also include the development stage, and finally the stage of post development with the product launching (Rozenfeld et al., 2006 and Baxter, 1998). Within this context, a job is graphic design because it follows projectual methodology, which is the reason for the design. That is, it has to be designed in any way (Villas-Boas, 1998 *apud* Gruszynski, 2000).

However, the PDP will be presented bellow. The choice of this method occurs due to consider the last stage of post development, unusual in projective conventional methodologies (Munari, 2006 and Bonsiepe, 1986).

2.1 Product Development Process

The PDP consists of activities through which it seeks from the market needs and technological possibilities and restrictions, and considering the competitive strategies and product and company, reach the specific design of a product and its production process, so that manufacturing is able to produce it (Rozenfeld et al, 2006).

The model includes the pre-development stage, where is the phase of Product Strategic Planning and Project Planning. Also includes the Development stage with the structuring of Informational Design, Conceptual Design, Detail Design and Preparation for Production. And yet at this stage, finally there is the Post development, where there is the monitoring of the production process and product launching.

This study emphasizes the post development stage. At this stage there is the relationship between user and brand sense. During this period, brand management acts to stimulate the creation of an ideal world in the mind of the consumer, able to persuade him and induce him to purchase, with a stimulus that transcend rational thought.

It is noted here that the purchase of products tends to transcend the basic or physiological needs of individuals (Maslow, 1954). However, this occurs because the need for representation and differentiation of the individual through the social status and self-realization. Therefore, the relationship of the user with the product today, no longer only due to physiological or basic needs, so this relationship becomes a complex experiment with symbolic interaction.

Within this context, this approach reaches a large field of study for cognitive ergonomics, since it seeks to communication strategies that promote symbolic articulations in the interaction between user and brand in the construction of the product image.

3   DESIGN AND COLLABORATION

The contemporary design has a multiplicity of actors from different areas that relate formally or informally in the solution design, are graphic designers, product designers, fashion, services or management. In this scenario, the projective processes are structured in networks design. These networks are systemically work together with the collaboration of various professionals. They share ideas, knowledge, materials, confirm to product development, participate in testing, modeling and prototyping; join in the production process and discuss strategies for insertion of the product launching.

Recurrently, collaborative work in design has been approached on projects in the strategic, tactical and operational task. However, collaborative work can occur in post development, with the insertion of the product for sale. This becomes more evident if we consider the possibility that the user of this new product has to interpret the cognitive strategies adopted in the aesthetics of the product and communication management in order to collaborate in the construction of cognitive perception and symbolic. Thus, the user shares and interacts with the layout of the atmosphere set by the sensory branding in order to promote and sell the product.

To understand the collaborative design, restricting only to the development process, can be taken based on the analysis from some level of human labor generally accepted in the literature:

- The Strategic Level: identification of features of the problem projectual;
- The Tactical Level: definition of objectives to be achieved in the project;
- The Operational Level: implementation of actions defined in the earlier stages, that is, execution projectual.

The meanings of the word collaboration, as Heemann et al. (2008), are linked to the concepts of cooperation, teamwork, networks, sharing and interaction. Thus, the collaborative process requires the cooperation and interaction of individuals, information and tasks between professionals and within the design process.

Collaboration is important in all stages of the project until the post development. In the post development is the process of building the brand image in the mind of the consumer as well as user interaction with the product. At this stage significant sensory experiences can stimulate purchase.

4   SENSORY BRANDING

The brands reflect the needs of individuals, is an identification process, propose, process, characterize social behavior, psychological, physiological and consumption.

A brand, a product or service must have a feature that differentiates the consumer's mind (Knapp, 2002). Thus, a brand carries a lot more than the representative character or figurative, but the ability to thrill through a unique and intimate contact with the consumer, to promote sensory experiences.

Brands make promises to the market, to get the experience of consumers in order to establish his reputation. Once they establish a loving relationship, create your beliefs, they become symbols of trust, make history and wealth (Strunk, 2007).

Branding is a brand management. The branding strategies linked to brands can contribute meaning the construction of emotional and rational consumer. The individual in this atmosphere of the product, promoted by branding, can interact sensitively and have deep experience with the product.

Lindstrom (2007) talks about companies that work with a strong management process in sensory brand, they are: Mercedes Benz, Disney, Mars, American Express, Reuters, McDonald's, Pepsi, Kellogg's and Microsoft. The author argues that big brands like these sharpen cognitive and sensory aspects of touch, taste, smell, look and sound that give it life, and add value to their products. Thus, multisensory brands produce profound emotional experience which consumers and boost it to the consumer.

While Lindstrom (2007) shows that the use of five senses is very important in winning the consumer, it provides unusual experiences. The brands must be "5D", five dimensions, that is a brand must act in five levels sense of human beings. A brand should not only promote a product or service visually. It takes some difference as adding some sound, which can be a song, powerful words or symbols. The combination of visual and auditory factors cause an impact of $2+2 = 5$. It is important to sharpen other sensory channels, and cognitive connections through taste, touch and smell. This helps to increase the total impact (Lindstrom, 2007).

Therefore, the branding seeks to understand the brands reached the top, and because they are so different. For example, the visual identity of Google and its many aspects of search; experiences of welcoming environments created by Starbucks, or even interface design developed by Apple and its iPod. These companies found single places in the market, they provide innovative consumer experiences, sharpen the five sensory levels and became multisensory brands.

It is in the post development, which will be managed communication actions to publicize the new product. In this context, the sensory branding seeks to sharpen the user's cognition, from strategies that value different sensory channels in order to promote appropriate symbolic articulation. In this approach, the process of communication has a purpose of transforming a situation. There is no a communication innocent. So there are two types of processes: the persuasion and manipulation (Niemeyer, 2003).

The effects of meaning are the result of a communication process: everything that happens at the meeting of the interpreter with the message. The knowledge of the interpreter, its values and its culture enables the right mix of signs so that the communication objectives are achieved (Niemeyer, 2003).

In this context, the branding stimulates consumption behavior, but only after the consumer interaction with the atmosphere of the brand, which is building the product image.

Bonsiepe (1992) confirms this thought, when he presents the design process consists in an interaction design. In this sense, "Design is the domain in which to structure the interaction between user and product, to facilitate effective action. Industrial design is essentially interface design".

However, based on the actions of branding to optimize the interface between user and product, visual communication strategies are adopted, such as media planning, layout point of sale, product testing and other strategies proposed by the atmosphere of sensory brand in order to involve the new product.

5   RESULTS AND DISCUSSION

The relationship between user and brand postmodern reaches a vast field of research and exploratory analysis for cognitive ergonomics. Given this context, the process of collaboration in design is evident, but little explored during the post product development.

The branding is important because it builds a sensorial atmosphere, which helps in the articulation of the symbolic image that the consumer has a new product. Some criteria can optimize the sensory branding, in order to provide profound experiences between user and brand, to build the product image, providing an impact 2 +2 = 5 (Lindstrom, 2007).

Lindstrom discusses the brands that are at the top make use of these criteria to promote cognitive stimulation such as touch, taste, smell, look and sound. That is, sharpen different sensory channels, which adds a strong differentiation to their products. In this respect, the user interacts with the communication system of the brand and from their sensory experiences contribute to the attribution of meaning to the product. Thus, Bonsiepe complements the design is essentially a process of interfaces.

Given this approach, we present here, that the articulation of some cognitive and sensory aspects such as branding strategies, can contribute to brand management, in order to improve the communication effectiveness of the product.

Niemeyer suggests that there is no communication innocent. Thus, branding can manipulate communication strategies, which the user will interact. From this thought, this paper presents the resulting effect on the interface between user and product is the construction of symbolic perception of the product by the user, which can stimulate or not your purchase. Although the author suggests that the effects of meaning are the result of a communication process and the user's knowledge of their culture and experience can determine the right mix of signs so that the communication objectives are achieved. That is, not only depend on the branding strategies promoted, but also the user culture.

However, sensory strategies adopted by brand, from communication management can determine the level of interaction, sharing and collaboration in the relationship between user and brand so as to produce different experiences in the design of interfaces.

6   CONCLUSIONS

This paper agues from the study of the Unified Model of Reference for the Product Development (Rozenfeld et al, 2006), collaborative design concepts and sensory branding applied to post-product development, identified the relationship between brand and user, and sensory experiences promulgated by the atmosphere of the brand, able to persuade the consumer.

The results and discussions presented in this paper show that during the post-development projectual there is a strong process of interaction and cooperation from the user about the actions of the brand in the construction of symbolic articulation. The user contributes, interacts and interprets the new product, contributing to the effectiveness of the communication performance of the brand. Thus, the sensory branding has strong applicability because it articulates the sensory actions that will promote an atmosphere of meanings that can involve the user and to induce the purchase with a force that transcends rational thought.

However, the paper argued that the actions of a multisensory brand should enable it to act on five dimensions of the senses. It is recommended that future studies build an atmosphere of the mark to identify the intensity and level of interaction can be obtained. In this perspective, different approaches can discuss sensory strategies adopted to promote the product. The vast field of research in cognitive ergonomics open here, allows research to be developed to measure the levels of sensory activity of a mark, to know the degree of experience 5D.

REFERENCES

Baxter, M. 1998. Projeto de Produto: Guia prático para o desenvolvimento de novos produtos. São Paulo: Edgard Blücher.
Bonsiepe, G. 1992. North/South Environment, Design. In: InCA, San Francisco. Chapter of the Industrial Designers Society of América, August: 14-15, pp. 08-93.
Bonsiepe, G. et al. 1998. Metodologia Experimental: Desenho Industrial. Brasília: CNPq/Coordenação Editorial.
Heemann, A. et al. 2008. Compreendendo a Colaboração em Design de Produto. Curso Superior de Tecnologia em Design de Produto, Centro Federal de Educação Tecnológica de Santa Catarina. Brazil.
Knapp, D. E. 2002. BrandMindset, fixando a marca. Tradução de Eliane Escórcio. Rio de Janeiro: Qualitymark.
Lindstrom, M. 2007. Brand Sense: A marca multissensorial. São Paulo: Artmed.
Lobach, B. 2001. Design Industrial: bases para a configuração dos produtos. São Paulo: Edgard Blücher.
Maslow, Abraham. 1954. Motivation and Personality. New York: Harper e Row.
Munari, B. 2006. Design e Comunicação Visual. São Paulo: Livraria Martins Fontes Editora Ltda.
Niemeyer, L. 2003. Elementos da Semiótica Aplicados ao Design. Rio de Janeiro: 2AB.
Rozenfeld, H. et al. 2006. Gestão de desenvolvimento de produtos – uma referência para melhoria do processo. São Paulo: Saraiva.
Strunck, G. 2007. Como criar identidades visuais para marcas de sucesso. 3 ed. Rio de Janeiro: Rio Books.
Villas-Boas, A. 2001. O que é [e o que nunca foi] design gráfico. 4 ed. Rio de Janeiro, Editora 2AB.

Section II

Cognitive Engineering: Workload and Stress

CHAPTER 8

Cognitive Workload Analysis and Development of a Stress Index for Life Science Processes

Manida Swangnetr[1], Biwen Zhu[2], David Kaber[2], Kerstin Thurow[3], Norbert Stoll[3] & Regina Stoll[3]

[1] Back, Neck and Other Joint Pain Research Group, Department of Production Technology, Faculty of Technology
Khon Kaen University, Khon Kaen, 40002, Thailand
manida@kku.ac.th
[2]Edwards P. Fitts Department of Industrial and Systems Engineering
North Carolina State University, Raleigh, NC, 27695-7906, USA
[3]University of Rostock, Center for Life Science Automation
Rostock, 18119 Germany

ABSTRACT

In life science process laboratories, common procedures often include calibration of systems for amino acid concentration measurement. Such procedures involve lab preparation, pipetting, automated systems programming, and data analysis. These activities may pose high levels of stress for technicians and can lead to excessive mental fatigue and errors. There is a need to empirically define an acceptable level of workload (or an overload threshold) for technicians as a basis for determining appropriate distributions of work tasks among operators or work-rest schedules. We conducted a field investigation at the Center for Life Science Automation (CELISCA) at the University of Rostock (Germany). Three professional technicians completed a series of tasks as part of an amino acid calibration process. Task completion time and technician perceived workload ratings using the NASA Task Load Index (TLX) were collected. Results revealed activity types and task exposure time to be influential in perceived workload. Mental and temporal demand components of workload were found to be significant in overall workload across task types. A composite stress index (CSI) was prototyped incorporating both cognitive and temporal demand ratings and task exposure time. An inspection of the distribution of CSI values suggested the 67th percentile to be a potential workload overload threshold for rescheduling of technicians across processes and among activity types in order to prevent mental fatigue and errors.

Keywords: cognitive workload, stress index, work-rest schedule, life science processes

1 INTRODUCTION

The advanced automation environment in life science process laboratories can pose substantial mental workload demands for operators. In common procedures, such as systems calibration for amino acid measurement, laboratory technicians are required to perform activities, including lab preparation, pipetting, automated systems programming, data analysis, etc. Such activities induce different levels of workload and stress based on task difficulty, task importance, time pressure, level of automation, etc. Although automation may be designed with the objective of reducing operator workload in each task, it is typical that such systems target manual handling operations and, to a limited extent, cognitive tasks. Consequently, the automation merely leads to a role change for technicians in which they are required to perform more highly demanding cognitive activities throughout a typical 8-hour workday. This can create very high levels of stress even for experienced technicians and can lead to excessive mental fatigue and errors.

The objective of amino acid calibration process is to facilitate fast test sample preparation and analysis in order to meet the demands of a high-throughput screening (HTS) system. The process generates initial calibration data (trends) for known amino acids that can be verified. The data serves to ensure accuracy and reliability in fast measurements for unknown test samples. The process for calibration curve generation has to be done for different settings of a High-Performance Liquid Chromatography and Mass Spectrometry (HPLC-MS) system (e.g., different types and lengths of HPLC-columns, different solvent flows, etc.) and for different types of amino acids. Although the system settings, task time, and detailed procedure depend on the types of amino acids, a typical calibration process includes: 1) sample preparation and labeling; 2) pre-dilution of samples; 3) substrate weight measurements; 4) dissolution of reagents; 5) performance of calibration tests; 6) addition of solvents; 7) reaction processes (e.g., incubator operation); 8) post-dilution of samples; 9) analysis using HPLC-MS (autosampler system operation); and 10) evaluation of data and report preparation. In general, the calibration process requires very accurate and precise work. Any calibration curve must be highly reliable in order to facilitate subsequent HTS of hundreds or thousands of unknown samples in, for example, cellular effects. Consequently, calibration is a very stressful process for lab technicians.

In order to address high levels of cognitive stress experienced by lab technicians, there is a need to empirically define an acceptable level of workload (or an overload threshold) for technicians as a basis for determining appropriate distributions of work tasks among operators or work-rest schedules. Operator stress levels should be considered when performing task scheduling; however, there is lack of prior research on defining, prototyping and validating stress indices. The primary objective of the present study was to develop a composite stress index (CSI) based on technician workload responses to various life science process activities as well as task exposure levels.

1.1 Literature review on workload measurement

The NASA-Task Load indeX (TLX; Hart and Staveland, 1988) is a subjective workload assessment tool that yields an overall workload score based on a weighted average of ratings across multiple subscales. The subscales represent workload demand components with three scales addressing demands imposed on the individual (i.e., Mental, Physical and Temporal demands). Three additional scales address the interaction of the individual with task (i.e., Performance, Effort and Frustration) (NASA Ames Research Center, 2003). These demand components are initially ranked by subjects based on importance to task performance. Weighting factors are determined for each demand through pairwise comparisons and integrated with individual scale ratings to compute the overall workload score. Individual demand component analyses can also be conducted by simply using the demand ratings. Hart (2006) observed that the component ratings can help diagnosis the source of a workload or performance problems. Prior research (Hill et al., 1992) has demonstrated the TLX to be easy to use by subjects and to increase the likelihood of accurate assessments of actual mental and physical workload.

Although the TLX was initially developed for use in the aviation domain, it has been applied successfully in field research in industrial and lab settings (see Hart, 2006 for an extensive review). Prior research has also found TLX responses to be sensitive to different types of activities (e.g., Matthews and Campbell, 1998; Rubio et al., 2004). In general, an increase in task demands leads to increases in perceived workload (Haga et al., 2002; Young and Stanton, 2005); however, this may occur more rapidly in certain tasks than others.

One disadvantage of the TLX is that it is a summative technique with ratings collected after subjective task performance and, therefore, ratings may be bias due to subject recall of workload. Another issue is that the TLX does not incorporate task exposure time. Although only a few studies have examined changes in workload over the duration of task performance (Haga et al., 2002), the general finding has been that task duration does affect workload responses. Mental fatigue appears to accumulate rapidly under higher demand conditions, including task performance that requires high levels of concentration.

In the present study, the prototype CSI focused on a subset of the demand components addressed in the TLX methodology, including mental, physical and temporal demands (Hart and Staveland, 1988). This was primarily based on the nature of the work of lab technicians. There is little time for self-appraisal (or performance assessment) in such work and activities have few intense physical demands integrated with cognitive performance. Furthermore, the skill level of technicians is such that frustration with task procedures is limited.

1.2 Hypotheses

The type of activity, as part of the common calibration procedure, was expected to influence technician mental, physical and temporal workload ratings [Hypothesis (H)1]. Each of the selected components of the TLX was expected to be sensitive to the nature of the activity, for example, high perceived mental ratings associated with data analysis work.

Task exposure time was expected to influence overall workload ratings. It was hypothesized that there would be correlations between task exposure time and perceived workload (H2). In general, subjective ratings of workload were expected to increase as task time increased. However, this relationship was expected to be non-linear with the influence of time-on-task in workload decreasing over time. That is, changes in the workload response were expected to be less significant with longer exposure (H3). Lab technicians may adapt or become less sensitive to the affect of workload on performance after long periods of work at the same activity.

Based on H1, H2 and H3, it was also hypothesized that the distribution of the proposed CSI values (i.e., workload ratings x exposure time) would be non-normal (H4). In general, it was expected that there would be a lower likelihood of increases in stress experiences with extended work time; that is, the trend of the CSI was expected exhibit a "ceiling" effect over time.

2 METHODOLOGY

In this section, we briefly describe the approach used to breakdown the target calibration process and for task performance time and workload data collection.

2.1 Task Analysis

A hierarchical task analysis (HTA) (not included here) was prepared based on a typical calibration process. The HTA involved identifying the goals, plans, tasks, operations and information requirements of technicians in the process. Several actual life science lab technicians were shadowed and interviewed for this analysis. Based on the information from the HTA, the basic activities performed by technicians were categorized as follows:
1. Sample preparation and labeling;
2. Substrate weight measurements;
3. Addition of pipetting solvent, auxiliary solution and reagents followed by shaking of the sample mixture;
4. Material handling and loading into a system;

5. Preparation of files and data analysis; and
6. Idle time, including incubator operation and HPLC-MS autosampler system run-time.

All of these activities were expected to induce different levels of mental workload due to differences in task information processing requirements.

2.2 Study Design and Procedures

A field study was conducted at the Center for Life Science Automation at the University of Rostock (Germany). Three professional lab technicians were observed in the completion of a series of tasks as part of the calibration process. At the beginning of the process, participants were asked to complete the pairwise rankings of the workload demand components, including mental, physical and temporal using a form translated into German. The technicians selected those TLX subscales that consider to represent the most important contributors to workload in the calibration process. The technicians were then provided with a brief description of each of the six basic activity categories and identification of how they were associated with the steps in the calibration process. The technicians were asked to tell the observer when they completed each step of the calibration process. At the end of each step, the observer recorded the step time and participants were asked to rate the TLX demand components. The ratings were used as a basis for determining overall perceived workload. The technicians were not asked to rate tasks that were classified as idle time.

3 DATA ANALYSIS AND RESULTS

In this section, we identify the statistical methods used for analysis of the workload and task time data and present results. We also describe the relationship of perceived workload with task exposure and the distribution of the CSI values.

3.1 Effects of activity type on workload

An Analysis of variance (ANOVA) model was structured to identify any influence of activity type on technician workload ratings during the calibration process. Results revealed the activity type to be significant in overall workload scores ($F(4,23) = 18.63$, $p<0.0001$) (see Figure 1). Based on post-hoc tests using Tukey's method, more cognitively demanding activities (e.g., data analysis and pipetting) caused significantly higher ($p<0.05$) workload than physical activities (e.g., loading and labeling). These results were consistent with operator rankings of the importance of the various workload demand components. Across all participants, cognitive demand was perceived and rated as a higher determinant of overall workload than the temporal component, followed by physical demand.

Standard deviations of the demand component ratings across operators and activities were also compared. The mental demand revealed the highest degree of variability among the three subscales ($SD_{mental}=5.27$, $SD_{physical}=4.14$, $SD_{temporal}=4.46$). This result indicated that mental demand was most sensitive for revealing differences among activities, as compared with physical or temporal demands. Taken together, all these results indicate that perceived workload in the particular life science calibration task was more dependent upon cognitive aspects of task performance than physical.

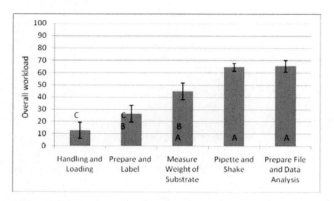

Figure 1 Overall workload means for basic activities and post-hoc test results based on Tukey's approach (means with different letter labels are significantly different from each other).

Additional ANOVAs were conducted to further investigate the effect of activity type on individual workload component ratings (i.e., physical, cognitive and temporal). Results showed that the activity category did not produce significant differences in operator perceived physical workload ($F(4,23)=1.62$, $p=0.201$) or cognitive workload ($F(4,23)=0.8$, $p=0.503$); however, significant differences in temporal workload were observed ($F(4,23)=6.08$, $p=0.0017$). In general, the pipetting task was less time demanding than the other activities and led to lower temporal demand ratings.

3.2 Relationship between workload and task time

Prior research found task exposure time to be correlated with perceived workload in an industrial context (Haga et al., 2002). On this basis, a correlation analysis was conducted to determine whether lab technician perceptions of mental, physical and temporal demand were related to activity time. A Pearson-product moment coefficient revealed a significant positive linear association of overall workload scores with calibration step times ($r=0.54$, $p=0.0018$).

To better understand the trend of perceived workload across task time, linear, exponential or logarithmic trends were fit to the overall workload scores per time. It was found that goodness-of-fit (R-square) was highest for the logarithmic trend, followed by a linear fit, and then an exponential fit ($R^2_{log}=0.39$; $R^2_{lin}=0.29$; $R^2_{exp}=0.23$). This fitting analysis suggested the relationship of perceived workload to task time was non-linear in nature. More specifically, the trend indicated the influence of time pressure or temporal demand was far greater for perceptions of overall workload for short-duration activities versus long.

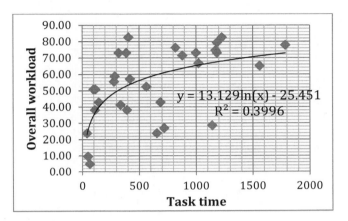

Figure 2 Logarithmic trend of workload scores against task time.

3.3 CSI prototype and distribution of values

Based on the above analysis, it was found that both activity time and the cognitive nature of the activity were influential factors in operator workload or stress levels. With this in mind, an initial candidate stress index was formulated as the product of a workload score, based on the cognitive and temporal demand components, with task time. (The physical demand component was disregarded, as it did not prove to be sensitive to the activity types.) Index values were determined for all observations on technician perceived workload and performance time for the various steps in the calibration process. Figure 3 presents the distribution of the CSI values and it can be seen that they closely conform with an exponential distribution. In specific, the stress index was found to dramatically increase above the 67th percentile.

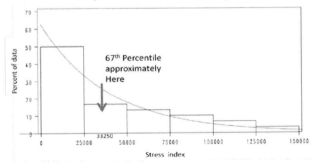

Figure 3 Histogram of values for proposed CSI; the fitted curve follows an exponential distribution with lambda=39897.

4 DISCUSSION

In line with our hypothesis (H1), there was a significant main effect of activity type on overall workload scores. Activities posing high cognitive demands, including data analysis and pipetting, caused significantly higher workload than physically intense activities (e.g., materials handling, loading and labeling). Results also revealed significant differences among activity types in terms of perceived temporal demand. Moreover, a high degree of variation in perceived cognitive load suggested mental demand ratings to be most sensitive for differentiating among types of life science tasks. These two components of workload (i.e., temporal and cognitive demands) were also considered by technicians to be more influential in overall workload across activities. These findings suggested that both cognitive and temporal demand ratings should be incorporated in any CSI.

In line with hypothesis (H2), there was a significant positive correlation found between task exposure time and workload ratings. Participants generally reported higher workload when working on longer tasks, which appeared to be due to perceived cognitive demands. This relationship also behaved in a non-linear manner (supporting H3) with a logarithmic trend providing the best fit. There appeared to be a plateau in perceptions of workload for longer activities; in this case, approximately greater than 13 min. The data indicated that temporal pressure was more of a critical factor in overall stress for lab technicians performing activities with a short duration. Related to this, like the overall workload score, temporal demand ratings were substantially correlated with task time ($r=0.45$).

The distribution of the proposed CSI values was also consistent with the hypothesis (H4). The PDF plot for the CSI revealed a skewed distribution, specifically exponential. From further inspection of the distribution using a CDF plot (see Figure 4 below), combinations of mental and temporal demand with task exposure time appeared to contribute relatively consistently to the likelihood of stress experiences up to the 67th percentile of CSI values. Beyond this percentile, there were dramatic increases in CSI values with reduced likelihood of occurrence. Therefore, such observations represented extreme cases of stress in the workplace with a high potential for mental fatigue and errors. Consequently, the 67th percentile of the CSI distribution may be considered a candidate for defining an overall workload or stress threshold for technicians in the calibration process.

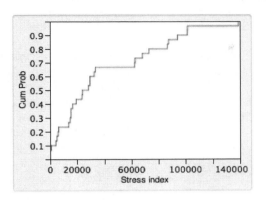

Figure 4 Cumulative density function plot for CSI values. (Note "knee-point" at 67th percentile.)

On this basis, CSI values observed for life science lab activities, which are above the 67th percentile (i.e., 33250), could be considered as a "trigger" for lab managers to reschedule technicians across processes and among types of tasks in order to reduce potentially problematic levels of workload and stress. Other solutions to managing technician stress level may include team performance of tasks or part-task performance among operators. It is also recommended that allotted task times be increased to reduce perceptions of time pressure (temporal demand) as far as possible while considering practical issues, such as the due date for a HTS job.

5   CONCLUSION AND FUTURE RESEARCH

The results of this study revealed life science lab activity types and task exposure time to be influential in technicians' perceived workload during common systems calibration procedures. Mental and temporal demands were found to have significant effects on overall workload across task types. A CSI was proposed integrating cognitive and temporal demands and task exposure time. Analysis of index values for a range of activities and times revealed a possible stress threshold that could be used as a basis for lab activity scheduling.

One limitation of the present study is that the CSI was developed based on a descriptive field study with three lab technicians. The study represents preliminary investigation on the development of a stress index. In general, additional data on a larger sample of subjects is needed to validate the findings. This study also examined only two factors in developing the CSI, including perceived workload and task exposure time. Other factors, such as work environment characteristics, schedule changes, etc. might affect technician stress levels.

Future research needs to expand the CSI and develop a quantitative approach to work task scheduling towards controlling worker stress levels. This may be accomplished by defining appropriate task mixes and work-rest schedules for the calibration process. One approach would be to incorporate the CSI, determined for common lab activities performed during an 8-hr work shift, as a constraint in existing heuristic scheduling algorithms (e.g., Shortest Processing Time, Earliest Due Date). Index values could be predetermined for each activity type. Any scheduling assignment could be constrained to ensure the mental workload of a technician does not exceed an upper-bound for the CSI, such as the 67th percentile. Beyond this, we plan to further develop the CSI by incorporating physiological responses, such as heart rate variability, along with the subjective ratings of mental and temporal demands. Such physiological measures have been demonstrated to be reliable indicators of cognitive workload (Wickens et al., 2004) and stress.

ACKNOWLEDGMENTS

We thank the University of Rostock for support of this research through a temporary Engineering Researcher position for Manida Swangnetr. We also thank CELISCA for access to high-throughput biological screening processes and allocation of biochemist time to the research effort. In particular, Dr. Dirk Gördes provided critical information on the HPLC-MS system and task analysis.

REFERENCES

Haga, S., Shinoda, H., and Kokubun, M. 2002. Effects of task difficulty and time-on-task on mental workload. *Japanese Psychological Research,* 44 (3): 134-143.

Hart, S. G. 2006. NASA-task load index (NASA-TLX); 20 years later. *Proceedings of Human Factors and Ergonomics Society Annual Meeting,* 50: 904-908.

Hart, S. G. and Staveland, L. E. 1988. Development of NASA-TLX (Task Load Index): Results of empirical and theoretical research. In *Human Mental Workload,* eds. P. A. Hancock and N. Meshkati. Amsterdam: North Holland Press.

Hill, S.G., Iavecchia, H.P., Byers, J.C., Bittner, A.C., Zaklad, A.L., and Christ, R.E. 1992. Comparison of four subjective workload rating scales. *Human Factors,* 34: 429-439.

Matthews, G. and Campbell, S. E. 1998. Task-induced stress and individual differences in coping. *Proceedings of Human Factors and Ergonomics Society Annual Meeting,* 42: 821- 825.

NASA Ames Research Center. 2003. *Instructions for the NASA TLX Version 2.0.* Moffett Field, CA: NASA.

Rubio, S., Diaz, E., Martin, J., and Puente, J. M. 2004. Evaluation of subjective mental workload: A comparison of SWAT, NASA-TLX, and Workload Profile, *Applied Psychology: An International Review,* 53(1): 61-86.

Wickens, C. D., Lee, J. D., Liu, Y. and Gordon-Becker, S. E. 2004. *An Introduction to Human Factors Engineering* (2nd Edition). NJ: Pearson.

Young, M. S. and Stanton, N. A. 2005. Mental workload. In *Handbook of Human Factors and Ergonomics Methods,* eds. N. Stanton, A. Hedge, H. Hendrick, K. Brookhuis, and E. Salas. Boca Raton: CRC Press.

CHAPTER 9

Interaction Effects of Physical and Mental Tasks on Auditory Attentional Resources

Abdulrahman Basahel, Mark Young and Marco Ajovalasit

Brunel University
Uxbridge, UK
Abdulrahman.basahel@brunel.ac.uk

ABSTRACT

Many tasks in the real world require simultaneous processing of mental information alongside physical activity. Whilst physical workload may have an impact on mental performance and attentional resources, most previous studies in ergonomics have not systematically investigated the impact of interactions between physical and mental demand on individual performance. The purpose of this study was to examine the influence of mental and physical workload interactions on auditory attentional resources and performance on verbal and spatial tasks. The moderate level of physical lifting improved the performance of auditory tasks under mental underlaod level. Surprisingly, rSO2 oxygenation changes were sensitive to changes in mental workload; also they were not sensitive to physical workload. The NIRS technique was valuable and sensitive in terms of reflecting the impact of mental and physical workloads on attentional resources and mental activities.

Keywords: physical workload; auditory mental workload; attentional resources; performance; brain activity

1 INTRODUCTION

Most researchers have focused on the examination of physical and mental demand impacts on individual's performance, separately. The mental workload has increased more than the physical workload in many jobs due to the rapid increase in technology in recent years; however, there are many jobs that require physical activities combined with mental tasks (e.g., firefighting, manufacturing, and assembly jobs) that can place stress on cognitive functions (Mozrall and Drury 1996). However, the authors who studied the effects of physical activity on cognitive function have found inconsistent and non-uniform results (Didomenico and Nussbaum 2008; Tomporowski 2003). Furthermore, researchers have focused on the effect of physical exercise on simple mental tasks (reaction time tasks) (Perry et al., 2008). According to some researchers, the optimum cognitive performance occurs under a medium level of physical exercise due to increased level of arousal (Audiffren et al., 2008). On the other hand, other researchers postulate that some levels of physical effort, such as a moderate level, can facilitate the mental process by increasing the percentage of blood flow, and thus oxygen, to the brain (Antunes et al., 2006), creating a defense toward any reduction in available oxygen in the brain due to mental stress and resulting in improvement in cognitive functions. Furthermore, Wilson and Russell (2003) said that the correlation between mental workload and performance is the same as between arousal level and performance, i.e., U-inverted. Therefore, it seems important to investigate the impacts of various physical and mental workload combinations on attentional resources performance.

NIRS, a recent method in neuroergonomics science, measures the impact of workload on brain activities (Parasuraman et al., 2008; Perry et al., 2009). Furthermore, the voluntary control of human actions or physical tasks combined with mental load (cognitive, perceptual, and affective processes) is one of the primary functions of the brain (Karwowski and Siemionow, 2003). As a result, the tasks, which include physical and mental demands, might place a heavy load on the brain function capacity of the operator. The purpose of this experiment is to examine the effects of physical lifting and mental workload interactions on auditory mental tasks (verbal and spatial).

2 METHODS

2.1 Design

The current study involves two experiments to investigate the effect of the interaction of physical workload (PWL) and mental workload (MWL) on individual attentional resources in the performance of auditory-verbal (arithmetic) and auditory-spatial (tone localization) tasks. The experiment used a 3×3 full factorial repeated measures design. Table 1 illustrates the nine conditions of interactions between the physical and mental arithmetic tasks. Table 4.2 presents the combination between physical demand and spatial figures at nine different levels.

Repeated measure analysis was used *within subject factors* (three physical and mental workload levels of interaction) and *between subject factors* (types of auditory mental tasks, i.e., verbal and spatial tasks).

Table 1 The nine conditions of interaction physical load and mental arithmetic tasks.

		Mental Auditory Verbal/Spatial Workload (MWL)		
		Low mental load	Medium mental load	High mental load
Physical Workload (PWL)	Low lifting load (8% body weight)	Performance will decrease under this condition due mental underload	Poor performance	Poor performance
	Medium lifting load (14% body weight)	Improves Performance	Acceptable performance	Poor performance
	High lifting load (20% body weight)	Poor performance	Poor performance	Performance will decrease under this conditions (overload)

The mental arithmetic tasks (auditory verbal task) included three levels of difficulty: low level (addition/subtraction numbers between 1 and 10), medium level (addition/subtraction problems with two numbers between 3 and 35) and high level (addition/subtraction problems with two numbers between 20 and 150 for the subtraction operation and between 20 and 105 for the addition). The tone localization tasks (auditory spatial task) included three levels of difficulty: low level (participants were asked to determine the sources of tones between two activated speakers placed at 270° and 30°), medium level (intermediate level, participants were asked to find the source of the tones from four speakers located in different positions (270°, 30°, 60° and 90°) and high level (difficult level, participants were asked to catch the auditory tone with eight speakers, (270, 30, 60, 90, 120 and 150 degrees). Physical lifting (8%, 14% and 20% of body weight) was as illustrated in Table 1; this physical task was similar to physical levels used by Didomenico and Nussbaum (2008).

2.2 Output measures

Dependent variables were included, namely: performance (number of correct responses, speed and accuracy), physiological indices (heart rate (HR) and regional cerebral oxygen saturation (rSO2)), and subjective assessments of overall workload (observed by using the NASA-TLX scores) (Hart and Staveland, 1988).

2.3 Particpants

The participants for each experiment were divided into two groups of 15 (one group for mental verbal task- experiment 1 and another for mental spatial task- experiment 2, in a between-subjects design), aged 25-35, with an equal balance of male and female participants in each group. In addition, the standard hearing test questionnaire (Self-Assessment of Communication, Schow and Nerbonne, 1982) was used to evaluate the hearing health of the participants; all participants had normal hearing. All subjects were healthy and did not have any history of back or musculoskeletal disorder. All participants were selected from the Brunel University staff and students.

2.4 Procedures

At the beginning, the participants were given a short introduction to the experiment in order to familiarize them with the steps. Also, the participants were provided instructions and advice on how to perform a lifting task and arithmetic mental task. The participants were then asked to affix the chest electrodes for the heart rate monitor on their chests so we could record the HR at baseline (rest) and measure the HR during the experimental conditions. The experiment included nine conditions, and counterbalancing between conditions was considered in order to reduce potential carryover effects and fatigue. Additionally, the NISR was fixed on the foreheads of the participants to measure the oxygenation (rSO2) of the brain.

At first, the participants' standing position was in the center of the experiment room; the dimensions of the chamber room were 3.30m×3.0m×2.68m this was used to isolate the outside noise. In addition, the room sound intensity level was nearly 33 dB(A) The participants were required to lift boxes from the floor to a 69-cm high table. They were required to do 4 lifts/min, and were free to select either of two methods of posture. The boxes were supported with two handles, and their dimensions were 35×35×30 cm. The table was placed in front of the subject and the boxes on the floor. The subjects were asked to keep their bodies and faces facing forward. The subjects were asked to lift the boxes onto the table and then place them back on the floor. Concurrently, they verbally answered the arithmetic problems given by the experimenter within an allotted fixed time (5 sec), to reduce the variation between each subject's speed in answering. The arithmetic sound was generated from the speaker at 70 dB (A) sound intensity (male voice).

Each volunteer completed nine conditions, and each condition was of 6 min duration. Also, each volunteer completed 25 questions at each level as accurately and quickly as possible in the allotted 6 minutes. The number of correct responses and the actual time required to complete the section was recorded directly by the software. Between each condition, each participant was given 5 min rest and the participant's NASA-TLX score was computed.

In the second experiment (i.e., tone localization task), the participants used the identical measure and equipment to the previous experiment. The tone generator software (NCH) was used to generate the tone and white-noise. The speakers were placed in the room at different positions and were each assigned a number. The speaker placed at 270° from the participant was assigned number 1; at 30°, number 2; at 60°, 3; at 90°, 4; at 120°, 5; and at 150°, 6. Two sounds were presented concurrently from two different speakers. For example, the pure tone and white noise were presented simultaneously from Speakers 1 and 2 for low auditory workload (two speakers active) with 400 ms duration. The participant responded to the pure tone concurrently with the lifting task and the experimenter recorded the answer in the program. Each condition level included 25 problems, and participants were given 6 minutes to complete each level. In addition, they took 5 minutes to rest and complete the NASA-TLX Score between each condition.

3 RESULTS

3.1 Performance

The ANOVA technique showed that the levels of difficulty of both factors separately—the physical and mental workloads—highly and significantly impacted the participants' answer accuracy in both visual mental tasks ($F(2,52) = 84.81$, $p<0.01$ and $F(2,52) = 57.21$, $p<0.01$, respectively). There was no interaction impact on accuracy ($p=0.332$). In addition, when the task levels (arithmetic and tone localization) increased, the accuracy decreased, except that the accuracy at low mental level and medium lifting load of interactions in both tasks was improved. In addition, there were no significant differences between both tasks under all levels of interactions ($p>0.05$)

According to contrast analysis, a significant difference was observed between all levels of physical workload ($p=0.027$). Additionally, the analysis showed a significant difference between mental workload levels in both tasks ($p=0.014$).

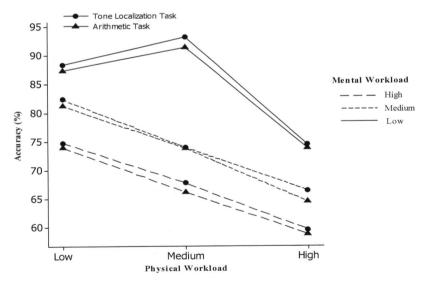

Figure 1 Means of accuracy of tone localization and arithmetic mental auditory tasks responses against physical and mental workload interaction.

The ANOVA technique showed that the mental workload factor highly and significantly impacted participants' time to provide correct responses ($F(2,52) = 4153.25$, $p<0.01$). In addition, the physical workload factor had a significant impact on participants' time to provide correct responses ($F(2,52) = 798.51$, $p<0.01$).

Moreover, the effects of physical and mental workload interactions on time for correct responses were significant ($p=0.01$). In addition, when the mental tasks levels (arithmetic and tone localization) and lifting weight increased, the average time for correct responses increased. The lowest time of correct responses appeared at medium lifting level × low mental demand of arithmetic and tone localization tasks. Auditory arithmetic while lifting consumed more time than the tone localization task. There were significant differences between tone localization tasks and arithmetic tasks at high level of mental load versus medium lifting load ($p=0.038$) and high mental load versus high lifting physical load ($p=0.031$), see Figure 2.

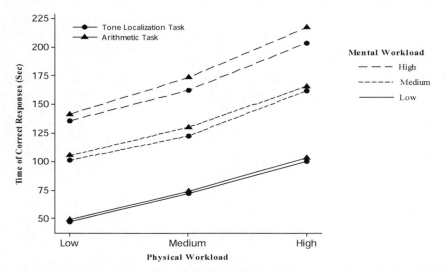

Figure 2 Means of correct responses time of tone localization and arithmetic mental auditory tasks responses against physical and mental workload interaction.

3.2 Physiological measures

There was a significant impact of mental workload on participants' HRs ($F(2,52) = 1210.02$, $p<0.01$) in both tasks. Furthermore, lifting workload levels in both auditory tasks (arithmetic and tone localization) had a significant effect on HRs of participants ($F(2,52) = 3120.51$, $p<0.01$). In contrast, the influence of lifting levels and mental workload tasks was not significant ($p>0.05$). Generally, the HRs' mean significantly increased when the physical and mental workload increased ($p<0.05$); see Figure 3. Moreover, the physical and mental workload interaction had not significant influence on HRs ($F(4,104) = 1.85$, $p=0.43$). Tone localization and physical lifting produced lower physiological stress than the arithmetic task condition, but there were no significant differences between both tasks ($p>0.05$).

As expected, the ANOVA technique showed that the mental workload factor highly and significantly influenced the participants' percentage of blood oxygenation in the frontal cortex of the brain (rSO2) ($F(2,52) = 153.86$, $p<0.001$). In addition, the physical workload factor had a significant impact on the percentage of oxygenation (F(2,52) = 59.82, p=0.012). Moreover, the effect of physical and mental workload interaction on rSO2 was not significant (p=0.63). However, the percentage of oxygenation in the brain increased when both mental task levels (arithmetic and tone localization) increased, whereas it decreased when the physical workload increased; see Figure 4. There were significant differences between both mental tasks at high mental workload while interacting with low, medium and high physical lifting tasks (p=0.013, p=0.022 and p=0.026, respectively).

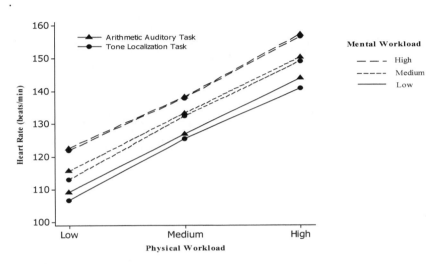

Figure 3 Heart rate means of tone localization and arithmetic mental auditory tasks responses against lifting boxes and mental workload

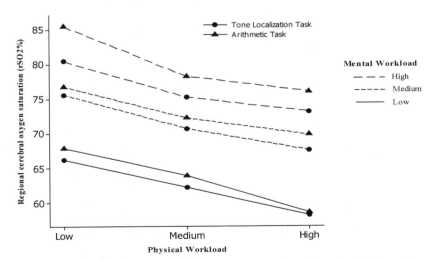

Figure 4 Brain oxygenation changes means of both tasks arithmetic and tone localization under nine levels of physical and mental workload interaction.

3.3 Subjective Assessment Tool (NASA-TLX)

The ANOVA technique showed that the mental workload factor highly and significantly impacted the NASA-TLX scores ($p<0.01$). In addition, the physical workload factor had a significant impact on the ratings ($p<0.01$). Moreover, the effects of the physical and mental workload interaction on NASA-TLX were not significant ($p=0.24$). The NASA-TLX score showed significant differences among mental workloads in both mental task levels and physical workloads ($p<0.05$). There were no significant differences between arithmetic and tone localization tasks among all levels of interactions ($p>0.05$).

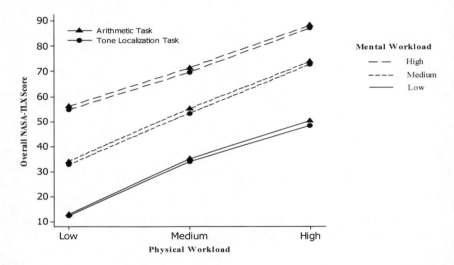

Figure 5 Overall workload assessment of arithmetic and tone localization mental task against physical and mental workload interactions, using the NASA-TLX rating.

4 DISCUSSION

The effects of physical and mental workload interactions on auditory tasks were investigated in this study. The findings of this experiment in terms of performance impact were that the accuracy and time of correct responses were impacted by the physical and mental workloads. Unexpectedly, optimum performance did not occur under medium level of physical and mental workload interactions in both auditory mental task conditions. Moreover, increased intensity in physical and mental workload (overload) led to poor accuracy and increased time of correct responses in arithmetic and tone localization tasks that increased stress on auditory attentional resources. That may be because the moderate level of arithmetic and tone localization tasks used in this study was complex for participants. However, the exciting result was that a moderate level of physical workload (14% of body mass) improved the cognitive information process under auditory mental underload level in tone localization and arithmetic tasks. Because the incremental increases in physical activity led to increased arousal level, performance increased. This is consistent with a previous study by Audiffren et al. (2008) that found that moderate cycling level improved performance of tone auditory reaction time tasks due to increased level of arousal. That is not consistent, however, with Didomenico and Nussbaum (2008), who found that increased levels of physical and mental workload did not impact arithmetic tasks performance. However, the differences between their study and current study may be because the auditory tasks that were used in the current study were more complex.

Heart rate was sensitive to physical and mental workload changes, since it increased significantly when workloads increased. In addition, a rise in rSO_2 percentage was associated with mental workload increases in both mental tasks. That means that increased auditory mental demands increased the percentage oxygenation changes in brain (brain activation), since increased brain activation indicate an imbalance between the amount of oxygen in the brain and the amount needed to meet high mental workload. That is consistent with Kikukawa et al. (2008), who found that increasing the level of mental demand in a pilot task (takeoff situation) leads to poor performance. On the other hand, the increase level of physical activity from low level to medium led to reduction the activation of brain since; the more of blood translated to brain so that the delivered oxygen to the brain was increased that reduced oxygenation changes percentages in the brain. As a result of that, it may the performance was facilitated at moderate physical level. This is similar to Antunes et al (2006) those suggested that the auditory cognitive functions supports by some physical exercise since, the physical activity makes a lot amount of blood flows to brain so, the amount oxygen increases which assess the mental information process. NASA-TLX increased significantly while physical and mental workloads increased and showed significant differences between physical levels and mental demands in both mental tasks conditions.

5 CONCLUSIONS

The balance between individual attention capacity and task workload was necessary, since some physical activity workload, such as moderate level, facilitates the low-level auditory cognitive information process and make it faster. In addition, high intensity levels of physical and mental workload led to poor responses. Increased physical activity did cause an increase in physiological arousal, which may improve performance. Significantly, an increased level of physical workload (to medium level) led to increased arousal level, which supported the cognitive task performance at low mental workload. Furthermore, this study supports the use of the NIRS technique that reflects the effects of physical and mental workload on brain activity and presents the percentage of oxygenation changes in the brain during workload interactions.

ACKNOWLEDGMENTS

I am greatly indebted to my supervisor, Mark Young, for his valuable comments, advice and support during this experimental study. I would like to express my thanks to second supervisor Marco Ajovalasit for giving me the comments.

REFERENCES

Antunes, H. K., Santos, R. F., Cassilhas, R., Santos, R., Bueno, O. and Mello, M. 2006. Reviewing on physical exercise and the cognitive function. *Rev Bras Med Esporte* 12: 97a-103e

Audiffren, M., Tomporowski, P. and Zagrodnik, J. 2008. Acute aerobic exercise and information processing: Energizing motor processes during choice reaction time task. *Acta Psychologia* 129:410-419.

Didomenico A. and Nussbaum M. A. 2008. Interactive effects of physical and mental workload on subjective workload assessment. *International Journal of Industrial Ergonomics* 38: 977-983.

Karwowski, W., and Siemionow, W. 2003. Physical neuroergonomics: the human brain in control of physical work activities , *Theoretical Issues In Ergonomics Science* 4: 175-199.

Mozrall, J. R. and C. G. Drury 1996. Effects of physical exertion on task performance in modern manufacturing: a taxonomy, a review, and a model. *Ergonomics* 39: 1179-1213.

Parasuraman, R. 2008. Butting the brain to work: Neuroergonomics past, present, and future, *Human Factors* 50: 468-474.

Perry, S., Thedon, T., and Rupp, T. 2009. NIRS in ergonomics: its application in industry for promotion of health and human performance at work, *International Journal of Industrial Ergonomics* 1-5.

Tomporowski, P.D. 2003. Effects of bouts of exercise on cognition. *Acta Psychological* 112: 297-324.

Wilson, G. and Russell, C. 2003. Real-time assessment of mental workload using psychophysiological measures and artificial neural networks. *Human Factors* 45:635-643.

CHAPTER 10

Applicability of Situation Awareness and Workload Metrics for Use in Assessing Nuclear Power Plant Designs

Svyatoslav Guznov[1], Lauren Reinerman-Jones[1], and Julie Marble[2]

[1] University of Central Florida (UCF), Institute for Simulation and Training (IST)
Orlando, USA
sguznov@ist.ucf.edu

[2] Nuclear Regulatory Commission (NRC)
Washington, D.C.

ABSTRACT

NUREG-0711 was developed by the Nuclear Regulatory Commission (NRC) to provide human factors engineering (HFE) guidance for conducting safety reviews of the HFE design development process in the evaluation of new and modified Nuclear Power Plant (NPP) designs. NUREG-0711 indicates that applicants should assess workload (WL) and situation awareness (SA) during the Verification and Validation of the design using state-of-the-art metrics for WL and SA. The majority of the metrics for assessment of WL and SA were designed for domains other than the NPP domain. Consequently, these metrics may need to be modified in order to be suitable for use in the assessment of NPP designs. Given the large number and variety of metrics, a tool that assists reviewers in determining the appropriate use of a given metric, is needed. In order to accomplish this, characteristics of WL and SA metrics that are particularly important to the NPP domain were determined. Next, a literature review was completed to identify available metrics of these cognitive constructs. Several metrics most applicable to the domain were identified. Implications and future directions are discussed.

Keywords: metrics, NPP, NUREG-0711, situation awareness, workload

1 NUCLEAR POWER PLANT DOMAIN

In order to provide reasonable assurance of safety, Nuclear Power Plant (NPP) designs are assessed by the US NRC at multiple stages and engineering elements. NUREG-0711 is used to guide the Human Factors engineering review of the main control room, as well as local control stations. Crews in NPP control rooms perform significant monitoring of systems while on shift. The operators are also required to follow standard procedures during normal, off-normal and emergency events in a highly coordinated manner. Human Factors Engineering is particularly concerned with the human performance component of NPP control. Human factors best practices indicate that human factors should be considered at the inception of the design. When dealing with a system as large and complex as a nuclear power plant, assessing the human factors of a finalized design is not cost effective nor would it lead to reasonable assurance of safety because of the interaction between design elements. Therefore, the review of the human factors begins at the programmatic level via a review of the plan to develop a human factors program. The Nuclear Regulatory Commission (NRC) created high-level guidance in NUREG-0711 Rev. 2 (O'Hara, Higgins, Persensky, Lewis, and Bongarra, 2004; Rev. 3 is in development), which provides guidance on the development of the Human Factors program and plan, and more detailed guidance on the development of human system interfaces (HSI) in NUREG-0700 Rev. 2 (O'Hara, Brown, Lewis, and Persensky, 2002); both are used to direct the reviewers in the examination of applicants' designs.

2 CURRENT GUIDELINES FOR EVALUATING NPP DESIGNS

NUREG-0711 provides reviewers with a process that defines multiple review elements that should provide feedback and feedforward of information during design and which gives guidance for the review of the process for verification and validation (V&V) of the design of the NPP control room. This guidance indicates that a V&V plan should be developed to assess the WL and SA of operators using the HSI and the interaction within the entire NPP system during the V&V element. V&V assessment of WL and SA should be performed using state-of-the-art metrics that are accepted within the HFE community. However, many of the existing metrics have been designed to assess WL and SA of tasks outside of the NPP domain (e.g., aviation, air traffic control, etc.). While the tasks described in these domains rely on similar cognitive functions that are used in the control of NPPs, the tasks in the domains themselves differ in terms of time available to perform, and reliance on symptom-based procedures (in the NPP domain) as opposed to event-based procedures or no formal procedures at all. Further, NPPs are controlled by a team of operators or crews, rather than single operators who perform individually; these crews are trained to very high standards using dynamic full-scope simulators. This training means that if a scenario is paused to apply a metric, it is possible the pause and subsequent questioning will influence the crew's thought process affecting the diagnosis and review of the current symptoms and state of the plant. In other words, if a metric administration allows the crew time to review (which might not normally be available during the scenario), the performance of the crew could change.

All of these characteristics mean that metrics for WL and SA should be adapted for application to NPP scenarios. The guidance provided by NUREG-0711 Rev. (2) specifies the cognitive constructs to be assessed; however, it does not and cannot prescribe how those assessments should be performed. There is a wide range of metrics that currently exist to assess WL and SA all having advantages and limitations. A tool that would ease the time for processing applications, assist in the development of requests for additional information, and assessment of the selection of metrics during V&V would be beneficial to NRC license reviewers. The tool under development will support the (1) determination of the applicability of a given metric for the NPP domain, and (2) determination of whether the metric is used in a manner that allows capturing WL and SA from the validation protocol and would be beneficial to NRC license reviewers. In addition, this work will be used by NRC staff in support of a new revision of NUREG-0711, including development of a companion document linking to more detailed information and tools for reviewers.

3 WORKLOAD AND SITUATION AWARENESS IN THE NPP DOMAIN

A first step to providing recommendations for reviewers regarding metrics for assessing WL and SA is to provide operational definitions. Those definitions lay the foundation of understanding for reviewers. This is then layered with a common ground of relevance to NPPs.

Various definitions of WL exist. NUREG-0711, Rev. 2, defines WL in terms of the cognitive and physical demands on the operator, and indicates that high, low and varying WL conditions should be assessed during V&V. WL is often defined as a balance between the amount of resources possessed by the human and the demand posed by the task (Veltman and Gaillard, 1996). Highly demanding tasks coupled with limited available cognitive resources result in elevated WL. Some models include elements that are not necessarily cognitive (such as time) as limited resources. Operator WL is affected by task structure, performance requirements, system design, and human information-processing limitations (Huey and Wickens, 1993).

Task structure requires the operators to perform various activities at NPPs. Typically, operators are required to monitor multiple displays for extended periods of time, while performing other control, administrative, and interface management tasks and sustaining awareness of the maintenance state of equipment. Prolonged monitoring can be classified as a vigilance task, which is highly demanding in terms of attentional resource usage, and, therefore, high in WL. WL during NPP control can be highly variable. During off-normal events the operators are required to multitask. Off-normal events, though extremely rare, can also be high in WL because the operators need to divide their attention among the tasks. During normal plant operation, low workload conditions are not infrequent in NPP operation during hours when little occurs.

Similar to task structure, performance requirements influence WL. NUREG-0711 provides guidance that HSI of NPP control panels should be designed to reduce uncertainty by providing operators with clear information about the state of the plant, thus reducing WL and enhancing SA. Task demand, which refers to the requirements of a task, is also decreased if the HSI affords operators the ability to execute necessary actions in a timely manner. In contrast to system design considerations, WL is also dependent upon factors intrinsic to the operators. Humans possess limited processing resources. Operators come to the task environment each day with a given amount of resources and this pool is affected by various factors including stress, fatigue, and emotion. These factors and more affect WL by detracting or enabling the allocation of resources for NPP control.

Thus far, the discussion has focused on individual operators and the way in which the WL of each plays a role in plant safety. However, NPP control is accomplished by a team of operators and therefore a definition encapsulating this aspect of WL is appropriate. Team WL refers to a combination of both taskwork and teamwork components (Cannon-Bowers and Salas, 1997). In the NPP domain, taskwork is the requirement of each crew member to accomplish a set of tasks (e.g., gauge monitoring). Teamwork requires each crew member to communicate information to appropriate team members, coordinate mutual activities, and provide peer checking. It is important to ensure that NPP teams have balanced levels of WL, meaning team members are not underloaded or overloaded.

WL is informative about the operator's evaluation of task difficulty, but SA provides insight into whether the operator observed all components of the task. In simple terms, SA is defined as understanding what is going on in the system (Endsley, 1995). Endsley identifies three levels of SA including operators' level of awareness and attention allocated to the current state of the system, understanding of the significance and importance of the system state, and the ability to infer about its future state. If the operators have good understanding of the processes happening at the plant, it is expected that they also have high SA.

As mentioned above, the majority of activities in NPP main control rooms are conducted in teams. In order to successfully accomplish the task, crew members need to have a clear picture of the processes occurring at the plant or have high individual SA levels. Team tasks are more likely to be successfully accomplished when every team member has high SA, collectively called team SA. This is an important concept to NPP operation, however, the topic of team SA is relatively new and unexplored.

4 SOLUTIONS TO CURRENT REVIEW LIMITATIONS

The metrics for assessing WL and SA are equally as numerous as the definitions for operationalizing these cognitive constructs. Therefore, an extensive literature review was conducted to create a database of existing metrics of WL and SA. The metrics were defined in the database by several properties and by characteristics of the experiment for which the metrics were used. Psychometric properties included reliability, validity, sensitivity, diagnosticity, and intrusiveness. Reviewers should look for metrics that show sufficient evidence for these characteristics. For example, a metric with low reliability would not produce consistent measurement of the characteristic of interest. A metric with a high level of intrusiveness, can disrupt primary task performance and be an additional source of WL. Characteristics of the experiments for which the metrics were used included sample size, sample type, domain, and environment/simulator fidelity. This information is useful for reviewers because they can readily see whether a metric is relevant to the NPP domain, real crews were used, or if a full-scale NPP simulator was employed. Based on these parameters, it becomes possible for a reviewer to assess the suitability of the for the NPP domain, and to form requests for additional information on the utilization of the metric in the design application.

4.1 Workload Metrics

The NASA-Task Load Index (TLX; Hart and Staveland, 1988) is a widely used metric of WL. It consists of six WL subscales including mental, physical, temporal, effort, performance, and frustration. The NASA-TLX has been used in the variety of domains including air traffic control, aircraft piloting, and, more importantly, NPP domain. The NASA-TLX is a multidimensional WL metric and, therefore, it allows discriminating various sources of WL. While the NASA-TLX is a well-known and widely used metric of WL, it is typically administered after the task or set of tasks. This limitation prevents its usage for scenarios with multiple events because these events cannot be assessed independently without pausing the task.

In order to overcome task intrusiveness, a short, single-dimensional WL metric can be used. The Instantaneous Self-Assessment (ISA; Tattersall and Foord, 1996) is a uni-dimensional WL metric that can be administered in parallel with the task. The participants are asked to make responses on a 5-point scale. Verbal responses to the scale can be used to minimize task intrusiveness even further. The limitation of this measure is that it has low diagnosticity, meaning it does not provide scores for different sources of WL.

Eye tracking and electroencephalogram (EEG) methods may be suitable to the NPP domain; however they have yet to be applied and accepted as metrics of WL for this particular domain. Two practical advantages of modern eye trackers are that 360 degrees visual view can be covered and head-mounted displays are not required. Eye tracking records various eye movements including blink frequency, pupil diameter, and saccadic movements. These metrics of eye tracking can help designers to identify problematic interface areas. However, Nearest Neighbor Index

(NNI; Camilli, Terenzi, and Di Nocera, 2007) and Index of Cognitive Activity (ICA; Marshall, 2002) provide more direct measures of WL. NNI involves a calculation of fixation points to derive randomness of scan patterns, such that the more random the pattern the higher the WL. ICA is a wavelet analysis of pupil diameter and provides a continuous scale of WL. That metric is most compatible with Seeing Machines FaceLab system. The limitation of this technology is that only one operator can be monitored at a time.

EEG records electrical activity occurring in the brain. EEG provides continuous data recording, which allows calculating several WL values throughout the task. Until recently, the biggest limitation of EEG method has been its physical intrusiveness to the task. The electrodes from the EEG cap were physically connected to stationary, bulky equipment restricting participants' movements. Modern EEG systems, such as Advanced Brain Monitoring's (ABM) X10 unit, can provide wireless and lightweight capabilities. A limitation of using EEG is that personnel must be extensively trained to collect accurate data. Standardization in experimental procedures is particularly important.

Since NPP operation is accomplished in crews, it is appropriate to measure WL in teams. Most team WL metrics use assessments created for measuring WL in individuals. Unfortunately, this method creates multiple problems such that it is unknown whether the psychometric properties are affected. Another serious problem is scoring and aggregating individual scores into one global WL score because it is unclear whether the WL scores need to be summed, averaged, or lowest or highest team member used. One of the most recent WL measures includes Team Workload Assessment (TWA; Lin, Hsieh, Tsai, Yang, and Yenn, 2011). This metric includes the NASA-TLX WL dimensions as well as team WL dimensions (e.g., communication, mutual back-up, and leadership). Overall, it appears that further development is needed to determine adequate use of team WL metrics in the NPP domain.

4.2 Situation Awareness Metrics

SA is another important construct to measure in NPP operators. The Situation Awareness Control Room Inventory (SACRI; Hogg, Folleso, Strand-Volden, and Torralba, 1995) is a self-report metric of SA. This metric is based on the SAGAT, which is a well-validated metric of SA (Endsley, 1987). In the SAGAT method, the simulator is frozen multiple times during the task. Operators are asked to answer a set of questions relevant to the state of the plant. The query administration is synchronized with the states occurring at the plant. Operator SA is judged by comparing operator understanding of the states of the NPP to what actually happens. The limitation of SACRI is that a new set of questions is needed for each scenario and is also intrusive because the task must be paused for administration.

The Situation Awareness Rating Technique (SART; Taylor, 1990) is a SA rating scale that is administered after the task. SART consists of three or ten dimensions depending on the level of resolution required. The three dimensional SART consists of: demand on attentional resources; supply of attentional resources; and understanding of the situation. Typically, the three-dimensional SART is sufficient in answering the question about perceived SA. The main benefit of SART is that it is administered after the task and is non-intrusive. However, as with most post-task metrics, SART provides just a summary of perceived SA during the task and might be subject to operator bias and memory decay. SART is most suitable when used to assess overall perceived understanding of the system.

Few team SA metrics currently exist and the application of the metrics is similar to those of team WL metrics. Therefore, similar problems are incurred. Even so, one metric can be discussed. Computerized Adaptive Rating Scale (CARS; McGuinness, 1999) is a subjective metric of SA, assessing perception, comprehension, projection, and integration of components of the environment. It is administered after the task and, therefore, is non-intrusive. This metric has been primarily used in a military domain, but has not been tested on its applicability to the NPP domain.

5 CONCLUSIONS AND FUTURE DIRECTIONS

The above consideration for metrics assessing WL and SA are relevant to the review of V&V programs for NPP designs and modifications. This is a critical element in the overall design process for a new NPP design. Note that the literature review and metrics determined to be most useful for V&V was extensive, but additional applications and needs still exist.

Metric utility is dependent upon relevant and up-to-date knowledge. New metrics of team WL and SA are anticipated to be created following the advances in team WL and SA theory. Few NPP domain specific metrics currently exist, hence it would be advantageous for more metrics to be tested and validated within the NPP domain. These metrics might already exist or one might be constructed to target the specific complexity of NPPs, but ultimately validation is required. In any case, progress has been made toward that end through the process described in this paper.

ACKNOWLEDGMENTS

This work was supported by the Nuclear Regulatory Commission (NRC) Nuclear Reactor Office (NRO) (JCN# N6768). The views and conclusions contained in this document are those of the authors and should not be interpreted as representing the official policies, either expressed or implied, of the NRC or the US Government. The US Government is authorized to reproduce and distribute reprints for Government purposes notwithstanding any copyright notation hereon.

REFERENCES

Camilli, M., M. Terenzi, and F. Di Nocera. 2007. Concurrent validity of an ocular measure of mental workload. In D. de Waard, G.R.J. Hockey, P. Nickel, and K.A. Brookhuis

(Eds.), *Human Factors Issues in Complex System Performance* (pp. 117–129). Maastricht, the Netherlands: Shaker Publishing.

Cannon-Bowers, J. A., and E., Salas. 1997. Teamwork competencies: The interaction of team member knowledge, skills, and attitudes. In: O'Neil, H. F. *Workforce Readiness: Competencies and Assessment* (pp. 151–74). Mahwah, NJ: Erlbaum.

Endsley, M.R. 1987. SAGAT: A Methodology for the Measurement of Situation Awareness. Hawthorne, CA: Northrop Corporation. NOR DOC 87-83.

Endsley, M. R. 1995. Toward a theory of situation awareness in dynamic systems. *Human Factors*. 37: 32–64.

Hart, S. G., and L. E., Staveland. 1988. Development of a multi-dimensional workload scale: Results of empirical and theoretical research. In: P. A. Hancock and N. Meshkati. *Human Mental Workload* (pp. 139–183). Amsterdam: North-Holland.

Hogg, D. N., K., Folleso, F., Strand-Volden, and B. Torralba. 1995. Development of a situation awareness measure to evaluate advanced alarm systems in nuclear power plant control rooms. *Ergonomics*. 38: 2394–2413.

Huey, B. M., and C. D. Wickens. 1993. *Workload Transition: Implications for Individual and Team Performance*. Washington, DC: National Academy Press.

Lin, C. J., T., Hsieh, P., Tsai, C, Yang, and T. Yenn. 2011. Development of a team workload assessment technique for the main control room of advanced nuclear power plants. *Human Factors and Ergonomics in Manufacturing and Service Industries*. 21: 397–411.

Marshall, S. P. 2002. *The index of cognitive activity: Measuring cognitive workload*. Proceedings of the IEEE Conference on Human Factors and Power Plants. Scottsdale, AZ.

McGuinness, B. 1999. *Situational awareness and the CREW awareness rating scale (CARS)*. Proceeding of the Avionics Conference ERA. Heathrow.

O'Hara, J., Brown, W., Lewis, P., and J. Persensky. 2002. *Human-system Interface Design Review Guidelines* (NUREG-0700, Rev 2). Washington, D.C.: U.S. Nuclear Regulatory Commission.

O'Hara, J., J., Higgins, J., Persensky, P., Lewis, and J. Bongarra. 2004. *Human Factors Engineering Program Review Model*. (NUREG-0711, Rev. 2 and supplements). Washington, DC: U.S. Nuclear Regulatory Commission.

Tattersall, A. J., and P. S. Foord. 1996. An experimental evaluation of Instantaneous Self-Assessment as a measure of workload. *Ergonomics*. 39: 740–748.

Taylor, R. M. 1990. "Situational Awareness Rating Technique (SART): The development of a tool for aircrew systems design." Proceedings of the AGARD Conference. Copenhagen, Denmark.

Veltman, J. A., and A. W. K. Gaillard. 1996. Physiological indices of workload in a simulated flight task. *Biological Psychology*. 42: 323–342.

CHAPTER 11

Effect of Occupational Stress among Workshop Trainees

Bahador Keshvar, Matrebi Abdulrani

University Technology of Malaysia
Email address:Keshvari_Bahador@yahoo.com
Alternative email address:kbahador2@life.utm.my

ABSTRACT

In the field of Neuroergonomics, scientists have obtained conflicting results regarding the effect of stress in forming short-term memory versus impairing it. The current study focused on the effect of gender differences among workshop trainees in a recall task. To accomplish the experiment, 120 (non-smoking) participants were recruited, consisting of 60 males and 60 females. The experiment was conducted in a university library (single room) and mechanical engineering laboratory. The experiment involved three stages. The first and second stages were held in a non-stressful situation in the library and the third stage was held in a stressful-situation in the mechanical engineering laboratory. Energy expenditure was evaluated for a recall task in each stage. Specifically, heart rate was measured via "oximeter finger pulse" twice per stage, once before the task and the second during task performance. Each heart rate measurement took approximately thirty seconds. Nicotine rate was also measured in order to determine non-smoking subjects "smokerlyzer." Before conducting the experiment, a questionnaire was administered among students in order to determine an appropriate reward to motivate competition in the experiment.

The analysis of heart rate reactivity and recall task efficiency were accomplished through SPSS 18. The mean heart rate in base line and stress conditions per subject was calculated. Correlations and regression analysis were the methods of analysis.

Gender differences influenced heart rate reactivity during the mental task in baseline and stressful situations. Results demonstrated that males adapted faster than females in the stress situation. Results of the mental task also demonstrated stress decreased mental efficiency in females as compared to males.

Keywords: stress, workshop trainees

1   INTRODUCTION

The impact of stress on short-term memory was investigated, including the influence of internal stressors, such as time pressure, unpredictability, loss control, and threat, and external stressors, such as age, gender, and race (Gaillard, 2008).

Researchers in the field of psychoneuroendocrinology have made substantial progress over the last decades in trying to disentangle conditions and mechanisms underlying the impact of stress on memory. The relation between gender and stress reveals several conflicting outcomes; numerous authors have determined that women find themselves in stressful circumstances more often than men (Almeida & Kessler, 1998; McDonough & Walters, 2001). Other authors have suggested that it is possible that women appraise threatening events as more stressful than men do (Miller & Kirsch, 1987; Ptacek, Smith, & Zanas, 1992). Furthermore, women have been found to have more chronic stress than men (McDonough & Walters, 2001; Turner et al., 1995; Nolen-Hoeksema, Larson, & Grayson, 1999). They are exposed to more daily stress associated with their routine role and functioning (Kessler & McLeod, 1984). To date, little is known about gender differences in response to repeated exposures to a stressor. There is a considerable amount of evidence indicating that women are both more psychologically and physiologically reactive to stressors than men. This includes greater heart rate (HR; Kudielka et al., 2004; Labouvie-Vief et al., 2003; Smith et al., 1997; Stoney et al., 1987).

Current study examined the relation between HR reactivity and mental performance in non-stressful (baseline) and stressful situation (stress) conditions among male and female trainees.

2   METHOD
2.1   Tasks and subjects selection

A total of 120 subjects were randomly selected in the University Technology of Malaysia from engineering majors including mechanical, electrical, and chemical engineering. They included four nationalities and races - Malay (15 male and 15 female), Chinese (15 male and 15 female), Iranian (15 male and 15 female) and Black-African (15 male and 15 female).

To evaluate short-term memory, a recall task was applied. The recall task consisted of numbers, words, and combinations of them. In the current study, 204 words were considered, which were derived from four groups of positive, negative, neutral and arousing words (Kuhlmann et al., 2005), from which 48 words were selected and used in the recall task. The recall task involved; i) memorizing the words for two minutes; ii) retaining the words for thirty seconds, and iii) writing down the words that were remembered.

The experiment consisted of three stages. First and second stages were held in a non-stressful situation (baseline) and the third stage was held in a stressful situation (stress). 24 of 48 words were applied at the first stage. The rest of the words (24 remaining words) were applied at the second stage. Combinations of words in the first and second stages made up the third stage words. There were six steps in the experiment; First Step, participant was asked to relax. The participant had some activity, such as walking, running, talking, and etc. before attending the experiment. Relaxation time was selected as 2 minutes according to the autonomic cardiac function measured. In the second step, HR reactivity was measured by having participants place their finger in an oximeter device and getting their "oximeter pluse" (HR1). The researcher registered all HR numbers, which were shown on a screen for 30 seconds (Mariaconsuelo et al., 2009). In step three, participants had 2 minutes to memorize 24 words. The 24 words were different at stage 1, stage 2, and stage 3. At step four, participants were asked to retain words they had memorized for 30 seconds. Measuring HR was repeated at this step (HR2). In step five, HR was measured by oximeter during 30 seconds. At step six, during 2 minutes participants wrote what they remembered. All steps were held in the second and third stages too.

2.2 Calculating recall task performance and heart rate

To calculate efficiency per stage, the number of words that participants remembered and wrote down was counted. The number was divided in 24 words (e.g., 14 words written to 24 words; 14/24=58.33% is efficiency of recall task per each stage).

To calculate efficiency of performance in the baseline, the mean efficiency of performance in the first and second stages was considered ($P_{BASELINE}$). P_{STRESS} determined efficiency of performance in the third stage. Differences between mean HRT in the first and second stages (baseline) and the third stage (stress) were determined in Tables 1and 2. The HR reactivity before the recall task (HR1) was calculated as

(Mean HR in baseline before recall task) − (Mean HR in stress before recall task) = Mean HR reactivity before recall task.

The HR reactivity during the recall task (HR2) was calculated as

(Mean HR in baseline during recall task) − (Mean HR in stress during recall task) = Mean HR reactivity during recall task.

2.3 Differences in stages

The third stage was different from the first and second stages. Some stressors, such as a new environment reward and counterfeit competitors, were added in the third stages. The first and second stages were held in a single-room at the library. The third stage was held in the mechanical lab with four counterfeit competitors.

In order to determine if subjects were non-smokers, a Smokerlyzer device was applied to measure nicotine rate in the body. Figure 1 shows 3 counterfeit competitors accompanying an actual participant in order to make a stress situation.

Figure 1 Actual and counterfeit participants in Mechanical Laboratory

A motivational factor (reward) could be earned by participants during the third stage. A questionnaire was distributed to determine a reward per race. Relation between student cost and reward was considered to find an optimum reward per race. Ten samples of data are shown in Tables 1 and 2. Heart rate in stressful situation and baseline, HR reactivity, and efficiency were determined for ten participants.

Table 1 Data collection for 10 samples before recall task

HR1 baseline	HR1 Stress	$P_{baseline} - P_{stress}$	Gender	$HR1_{baseline} - HR1_{stress}$
56.36	62.35	-12.00	Male	-5.99
62.15	69.25	-28.00	Male	-7.10
57.00	66.35	-10.00	Male	-9.35
59.21	77.31	-13.00	Male	-18.10
63.87	67.23	-12.00	Male	-3.36
63.58	79.23	-36.00	female	-15.65
58.14	64.68	-10.00	female	-6.54
64.52	74.23	-30.00	female	-9.71
65.17	71.52	-25.39	female	-6.35
58.14	64.68	-10.00	female	-6.54

Table 2 HR data for 10 samples during task and performance differences

HR2 baseline	HR2 Stress	P $_{baseline}$ − P $_{stress}$	Gender	HR2 $_{baseline}$ − HR2 $_{stress}$
69.24	71.02	-15.32	Male	-1.78
65.58	79.35	-26.34	Male	-13.77
65.10	82.35	-32.21	Female	-17.25
66.93	69.32	-10.00	Male	-2.39
66.32	75.23	-15.00	female	-8.91
68.79	81.23	-29.00	Male	-12.44
69.35	64.68	-10.00	female	4.67
65.08	75.36	-9.00	male	-10.28
70.15	77.25	-18.00	female	-7.10
58.14	64.68	-10.00	female	-6.54

3 ANALYSIS

Correlations between HR, performance, and gender were evaluated by SPSS 18. Correlations and Pearson coefficient were applied for p-value <0.05. Mean of P$_{BASELINE}$ was demonstrated performance in the first and second stages. P$_{STRESS}$ was demonstrated performance in the third stage.

Table 3 Relation between heart rate (HR1), performance and gender before task

Gender	Variables	Relationship	P-value
Male	(Mean of P$_{BASELINE}$) − P$_{STRESS}$	-0.21*	0.05
	(Mean of HR$_{BASELINE}$) − HR$_{STRESS}$		
Female	(Mean of P$_{BASELINE}$) − P$_{STRESS}$	-0.40**	0.001
	(Mean of HR$_{BASELINE}$) − HR$_{STRESS}$		

**Correlation was significant at 0.01 level (2-tailed). * Correlation was significant at 0.05 level (2-tailed).

Regression analysis results (see Table 4) indicated that the regression correlation coefficient between two variables (1- performance 2-Mean of HR reactivity) for males was 0.21 and for females was 0.40.

Table 4 Regression analysis results

P	T	β	R^2	R	P	F	Criterion variable	Predictor variable	Gender
0.05	-1.976	-0.17	0.06	0.21	0.05	3.904	($HR_{BASELINE} - HR_{STRESS}$)	performance	Male
0.001	-3.371	-0.22	0.16	0.40	0.001	11.364	($HR_{BASELINE} - HR_{STRESS}$)	performance	Female

The influence coefficient (β) in males revealed that when performance increased one-unit, ($HR_{BASELINE} - HR_{STRESS}$) decreased 0.17% and also in females when performance increased one unit, ($HR_{BASELINE} - HR_{STRESS}$) decreased 0.22%.

4 DISCUSSION

The present study examined the effect of gender on short-term memory. According to previous findings, stress influences short-term memory in females more than males. Stress also influences retrieval memory, which is more significant in females (Joels et al., 2006; Lupien & McEwen, 1997).

Stress has an impact on how much we remember (Joels et al., 2006; Kim & Diamond, 2002; Lupien & McEwen, 1997). Stress influences memory consolidation. Epinephrine and cortisol are hormones released by glands, which are located above the kidneys, and also result in releasing oxytocin. Oxytocin levels in females are more than in males. Available data concerning the effect of oxytocin on memory are often inconsistent. Females have the same response to stress as male, leaving them somewhat vulnerable to cortisol and adrenaline. But then female also secrete oxytocin from the pituitary gland, which helps scale back the production of cortisol and adrenaline and minimizes their harmful effects. Males also secrete oxytocin when they are under stress, but they produce it in amounts less than females do. It can be concluded that high amounts of oxytocin in females increase the effect of stress on recall task efficiency among females as compared to males. Based on physiology findings, Stroud (2002) determined that women show greater cortisol reactivity than men. In addition, Kirschbaum (1996) found a strong correlation detecting a more pronounced cortisol response was associated with poorer memory. The results confirmed pervious findings of researchers such Kuhlmann (2005). The results of the current study found the efficiency of mental tasks and physical tasks in stressful situations was more degraded than in baseline situations. The findings confirmed the findings of researchers such as Stroud (2002) and Kirschbaum (1996), revealing that gender differences influence recall task efficiency in stress and baseline situations. Results determined female efficiency is less than male efficiency in stressful situations. Our findings confirm the pervious findings. It also demonstrated males adapted in stressful situations faster than females. The current study determined that memory consolidation in men may be due to increased efficiency in stressful situations.

5 CONCLUSION

The findings of the experiment showed the effect of stress on mental performance is more significant in females as compared to males. Results demonstrated that males adapted faster than females in stress situations. Results of the mental task also demonstrated the effect of stress decreased mental efficiency in females more than males.

6 FUTURE WORK

The project results could be useful in the following areas:

i) The experimental methodology could be used with different factors such as age and different nicotine-rate participants.

ii) The study was focused on 120 subjects. The number of subjects in this experiment could serve as a sample of a larger number of subjects (thousands) in future studies.

iii) Neuroscientists could apply the results in the field of "perceptual process" in their research.

iv) The current experiment was not limited to special groups, such as small or huge companies. It encompasses all companies for which gender differences interact with workload. Stressors could influence efficiency such as fatigue, work-overload and time pressure. Based on the research findings, gender difference should be considered when designing workload. Short-term memory tasks show the effect of stress is significantly different between males and females. Human resource managers could use the results in order to choose the fittest candidate among those who volunteer for tasks with high levels of occupational stress.

REFERENCE

Almeida, D. M., & Kessler, R. C. (1998). Everyday stressors and gender differences in daily distress. Journal of Personality and Social Psychology, 75, 670–680.

Gaillard, A.W.K. (2008). Concentration, stress and performance. In P.A. Hancock & J.L. Szalma (Eds.), *Performance under stress* (pp. 59-75). Hampshire, England: Ashgate Publishing.

D.J. Kennett, Devlin, M.C. & Ferrier, B.M. (1982). Influence of oxytocin on human memory processes: Validation by a control study. *Life Science*, 31(3), 273-275.

Engelmann, M., Ebner, K., Wotjak, C.T., & Landgraf, R. (1998). Endogenous oxytocin is involved in short-term olfactory memory in female rats. *Behavioural Brain Research*, 90(1), 89-94.

Joels, M., Pu, Z., Wiegert, O., Oitzl, M.S., & Krugers, H.J. (2006). Learning under stress: how does it work? *Trends in Cognitive Science*, 10, 152–158.

Kessler, R.C., & McLeod, J.D. (1984). Sex differences in vulnerability to undesirable life events. *American Sociological Review*, 49, 620–631.

Kudielka, B.M., Buske-Kirschbaum, A., Hellhammer, D.H., & Kirschbaum, C. (2004). Differential heart rate reactivity and recovery after psychosocial stress (TSST) in healthy children, younger adults, and elderly adults: the impact of age and gender. *International Journal of Behavioral Medicine*, 11, 116–121.

Kuhlmann, S., & Wolf, O.T. (2005). Cortisol and memory retrieval in women: influence of menstrual cycle and oral contraceptives. *Psychopharmacology* 183, 65–71.

Lupien, S.J., & McEwen, B.S. (1997). The acute effects of corticosteroids on cognition: integration of animal and human model studies. *Brain Res. – Brain Res. Rev.*, 24, 1–27.

Labouvie-Vief, G., Lumley, M.A., Jain, E., & Heinze, H. (2003). Age and gender differences in cardiac reactivity and subjective emotion responses to emotional autobiographicalmemories. *Emotion* 3, 115–126

McDonough, P., & Walters, W. (2001). Gender and health: reassessing patterns and explanations. *Social Science & Medicine*, 52, 547–559.

Miller, S. M., & Kirsch, N. (1987). Sex differences in cognitive coping with stress. In R. C. Barnett, L. Biener, & G. K. Baruch (Eds.), *Gender & Stress* (pp. 278–307). New York: The Free Press.

Nolen-Hoeksema, S., Larson, J., & Grayson, C. (1999). Explaining the gender differences in depressive symptoms. *Journal of Personality and Social Psychology*, 77, 1061–1072.

Ptacek, J. T., Smith, R. E., & Dodge, K. L. (1994). Gender differences in coping with stress: When stressors and appraisal do not differ. Personality and Social Psychology Bulletin, 20, 421–430.

Smith, T.W., Nealey, J.B., Kircher, J.C., & Limon, J.P. (1997). Social determinants of cardiovascular reactivity: effects of incentive to exert influence and evaluative threat. *Psychophysiology* 34, 65–73.

Stoney, C.M., Davis, M.C., Matthews, K.A. (1987). Sex differences in physiological responses to stress and in coronary heart disease: a causal link? *Psychophysiology*, 24, 127–131.

Valentini, M., & Parati, G. (2009). Variables influencing heart rate. *Progress in Cardiovascular Diseases*, 52, 11–19.

Section III

Cognitive Engineering: Activity Theory

CHAPTER 12

On the Relationship between External and Internal Components of Activity

W. Karwowski, F. Voskoboynikov, G. Bedny

University of Central Florida,
FL, USA
wkar@ucf.edu
Baltic Academy of Education,
St.-Petersburg, Russia
fredvosko@hotmail.com
Evolute,
Louisville. KY, USA
gbedny@optonline.net

ABSTRACT

Activity Theory (AT) had a long history in the former Soviet Union. Three prominent Russian scholars - Vygotsky, Leont'ev and Rubinshtein - were responsible for the development of general Activity Theory. AT was initially formulated by Rubinshtein and Leont'ev. The cultural-historical theory of development of human mind developed by Vygotsky was also critically important for AT. According to Vygotsky, external social activity is the source of internal mental activity. This idea was specifically formulated by him as the principal of *internalization*. Rubinshtein argued that individual psychological characteristics of human are not completely derived from the social environment. We will analyze from the general Activity Theory and the systemic-structural activity theory (SSAT) perspectives how the relationship between the external and internal components of activity affects the development of the human mind.

Keywords: activity theory, systemic-structural activity theory, external social activity, internal mental activity, internalization, cultural-historical theory.

1 INTRODUCTION

AT was initially formulated by Rubinshtein (1940) and Leont'ev (1978). They both proceeded from the general statement that object-practical activity is the principal source of mental development. However, their views on some other aspects of AT were significantly different. For example, Rubinshtein's major point was that "external causes act through internal conditions", whereas Leont'ev did not give sufficient value to the role of internal conditions in mental development. By Leont]ev, "external and internal activity have similar structure". The leading idea of Rubinshtein's concept was that human action changes not only the object but also the subject himself. Rubinshtein is considered as one of the main contributors into development of AT.

Vygotsky (1962) developed the cultural-historical theory of development of human mind. According to Vygotsky, external social activity is the source of internal mental activity. This idea was specifically formulated as the principal of *internalization*. Activity in his view was considered not so much on individual plane but more on the plane of social interaction. Rubinshtein disagreed with the concept of internalization as a driving force of mental development. He argued that a person does not simply internalize ready-made standards. His important assertion regarding mental development was that "external influences on mental development always act through internal conditions". That is, psychological characteristics of the individual are not completely derived from the social environment. The connection of personality with the social world can be understood as being mediated by activity. By Rubinshtein and Leont'ev, not words or social interaction, but interaction of subject with object played the leading role in mental development. Rubinshtein though paid insufficient attention to the Vygotsky's semiotic aspects of activity.

One should distinguish the external behavioral and internal mental actions. External actions include various motions and transform material or tangible objects. Internal actions transform images, concepts or propositions and non-verbal signs in the mind. We will follow up how both the external social and internal mental affects the development of human mind.

2 THE PERSONALITY PRINCIPAL IN THE STUDY OF HUMAN DEVELOPMENT

Rubinshtein was the first one who introduced *personality principal* in psychology that integrates individual and social aspects in the study of human development. According to this principal, human development is the results of interaction of material and social practice with human subjectivity. This principal eliminates the contradiction between social and intraindividual aspects of human development. Rubinshtein's personality principal addresses inadequacies in behaviorists' approach, which was introduced by Skinner (1974). In Skinner's view, human reactions are considered to be the results of external influences, ignoring the subjective aspects of reactions. By Skinner, external reality is a variety of stimuli to which a person must react, that is, human emerge as reactive organisms. Subjectivity is absent in his studies. Behaviorism ignores mediated functions of activity, which provide a basis for personal development. In AT, a person who interacts with a situation is considered the subject; that is, we talk about actions and cognition, not about reaction to stimuli.

Rubinshtein wrote that through the organization of the individual practice, society shapes the content of individual consciousness. His famous quotation "external acts through internal" emphasizes the dependence of activity on the subject's individual features. By Rubinshtein's personality principal, the psychological existence of human being is connected with real existence. The social aspect depends on the individual, just as individual depends on the social aspect. In the same social environment, different individuals act differently, and they are impacted by the social environment in different ways.

Both Rubinshtein and Leont'ev suggested that the leading role in the development of the human mind is not played by words or social interaction, but rather by the interaction of the subject and object. Rubinshtein was one of the first to formulate ideas on subject-object and subject-subject relationships in activity and the object-related character of activity. Nature exists independently of the subject; however, object can emerge only during the interaction with the subject.

3 THE INTERNALIZATION CONCEPT

The interrelationship between the external and internal components of activity takes a central place in Vygotsky's work. He attempted to track how the social interpenetrates individual consciousness. This idea was specifically formulated as the principal of *internalization*. According to Vygotsky, internalization is first of all a social process where "external" means "social". External social activity is the source of internal mental activity. Language, as semiotic aspect of internalization, becomes the major mediator between social and individual functioning; between external and internal activity. In Vygotsky's work, the development process is led by social interaction. While Vygotsky's theoretical statements played an important role in the development of AT, analysis of his theory demonstrates that activity was considered not so much in individual plane but more in the plane of social interaction. Rubinshtein and his followers criticized Vygotsky's work on the grounds that he did not consider the activity of individuals in the process of development of human mind sufficiently enough. We would also like to emphasize that activity is not purely adaptive; it is also transformative and creative (Bedny, Karwowski, and Voskoboynikov, 2010). The specificity of knowledge acquisition and socialization is dependent not only on social but on the specificity of individual activity as well.

The internalization concept is important for understanding the relationship between the external and internal processes and personal development processes. Leont'ev also considered the problem of external and internal aspects of activity through the lens of internalization. According to Leont'ev, the structure of object-oriented activity determines individual consciousness to a significant degree. Leont'ev wrote: "external and internal activities have similar structure", but he did not consider sufficiently enough the individual-specific regulative aspects of activity. Rather, he suggested that the structure of individual consciousness is predetermined by the structure of external object-oriented activity. In actuality, social determination of consciousness has its sources not only in "external influences", but also in "internal influences", which depend on the specificity of individual activity and methods of individual performance. According to the principle of self-regulation of activity, internal plane of activity is a result of active formation of internal mental actions and operations.

4   ACTIVITY AS A SELF-REGULATING SYSTEM

To understand the relationship between the external and internal components of activity, we should refer to the systemic-structural theory of activity (SSAT). By the systemic-structural approach, *activity is considered as a coherent system of internal mental processes and external behavioral processes and motivations that are combined and organized by mechanisms of self-regulation to achieve conscious goals.* The concept of self-regulation is critical for understanding activity as a system and for analysis of preferable strategies of performance. The SSAT concept of self-regulation is different from the concept of self-regulation outside of activity theory where models of self-regulation are homeostatic. Homeostatic relates to homeostasis which is the ability or tendency of an organism or a cell to maintain internal equilibrium by adjusting its physiological processes. The goal in homeostatic models of self-regulation is viewed as an unchangeable readymade standard. In reality, the goal is interpreted, accepted, or reformulated by a subject. In activity theory, self-regulation is a goal-directed process where the goal has an integrative systemic function. During the task performance, the goal becomes more specific and clear. At any point of time, cognitive processes are integrated to achieve a specific purpose of activity.

Self-regulation manifests itself through both non-conscious and conscious levels (Bedny and Karwowski, 2007). At the non-conscious level, conscious and verbalized aspects of self-regulation play a subordinate role, and this level is particularly important when imaginative and non-verbalized strategies of activity play the leading role. At the conscious level of self-regulation, verbal and logical aspects of activity are dominant. Both levels of self-regulation are interdependent and the relationship between them is dynamic. This interdependency gives rise to the formation of different strategies of activity, which are adequate to the external and internal conditions of activity. Learning is considered a self-regulating process during which strategies of activity are transformed.

At the unconscious level of self-regulation, condition unfolds as an uninterrupted process. Automotive mental operations are not organized into cognitive actions. This can be explained by the fact that the unconscious level of self-regulation is not subordinated to conscious goals. Activity is triggered automatically and performed through unconscious automotive reflective processes, and the subject is only conscious of the results of this process. The conscious level of self-regulation presents itself not only as a process but also as a system of logically organized actions. Each action is organized according to mechanisms of self-regulation and has a beginning and an end. At the conscious level of self-regulation, activity can be considered a hierarchically organized system of self-regulative stages of uninterrupted reflective processes. At the same time these processes are discrete. Therefore, at the conscious level, cognition is continuous and interrupted at the same time. Understanding the principal of self-regulation of activity helps us to understand the relationship between uninterrupted and interrupted aspects of self-regulation.

Understanding how activity is organized helps explicate the relationship between the external and internal components of activity. The socially determined aspects of our cognition are not only based on "external" influences, as suggested by the Vygotsky's cultural-historical theory of development. Nor do they wholly depend on object-oriental activity, as suggested by Leont'ev. Psychic activity emerges as a function of social existence of the individual, and, as a result, the ability for psychological reflection develops. Psychological reflection is not a passive mirror-like reflection; it possesses active features that imply some systems of mental stages and operations and is always organized as a self-regulation process. Since the process cannot be fully determined in advance, it contains situated elements that are developed during self-regulation process of reflection. The more complicated a person finds a task, the more important and complicated the process becomes. The most complicated reflective process is thinking.

According to the principal of reflection, psychic or cognition functions are organized as a self-regulation process. As the conscious level of self-regulation this process is organized as a system of logically interdependent cognitive action that are transformed one into another as activity unfolds. These actions include mental operations. At the unconscious level of self-regulation this process unfolds as automotive unconscious operations. Psychological determination does not depend on social or external factors only but also on internal influences derived from the mechanisms of self-regulation, which integrate external and internal components of human activity.

5 CONCEPT OF INTERNALIZATION IN SYSTEMIC-STRUCTURAL ACTIVITY THEORY

Leont'ev and Rubinshtein concentrated their efforts on the study of practical object-oriented activity as a source of mental development. Vygotsky mostly paid attention to the semiotic aspects of activity and their connections with the problem of social interaction. According to SSAT, the aspects of activity and the development of personality are interconnected. The objective world becomes open to humans only in the unity of the object-oriented and semiotic aspects of activity and social relations. Social relations and individual practical activity exist in unity. A child develops into a person through the process of individualizing himself from the surrounding environment. A child can become a person only in social environment because only humans can communicate the knowledge of how to utilize the artifacts of human culture.

The child is born into the objective world with already developed symbolic systems that correspond to this world. Human activity, from the very beginning, is both object-oriented and sign-oriented. Vygotsky, in his famous example interpreted the gesture of a child, which signifies the material act into a sign. If a child attempts to grasp an object unsuccessfully, mother realizing that the movement indicates something, immediately helps him. Thus, during the social interaction with the mother, the child also interacts with the objective world. Both aspects of activity are equally important. Gradually the child becomes aware that his hand movement becomes a sign, which can be understood by others, that is, the object-practical action is transferred into semiotic action. The internal world of humans cannot be portrayed as a simple result of transformation of the external into internal. It is rather the results of the formation of the construction of the internal based on the unitary process of social- and object-oriented interactions. Furthermore, it is important to note that during the development of sign systems, a child learns how to manipulate these systems in some degree independently based on individually developed strategies.

During the first year of mental development children acquire the meaning of different objects that have precisely assigned functional purposes. Later on, they gradually begin to acquire the meaning of things that do not have assigned purposes. The acquisition of meaning as an internal mental tool cannot be reduced to the manipulation with external object tool that has a particular purpose. It is not the process of transformation of external operations with external tools into internal operations with mental tools. Internal mental operations do not replicate external behavioral operations. Mental development is a result of interaction between external and internal operations based on the mechanism of self-regulation and, therefore, always includes some unique individual components.

At the beginning of a learning process, a learner manipulates signs and symbols by means of external objects using different instructions and schema, which facilitate externalization of internal mental activity. We can say that mental activity is guided by the external orienting components of activity. Therefore, external behavior and internal activity depend on each other. During feed-forward and feedback influences between external and internal components of activity, a mental plane of activity is developed. This leads us to conclude that mental activity cannot be considered the transformation process of external process into internal. Internal plane of activity can be significantly different from its external form. At this stage of performance, mental activity becomes relatively independent of external activity and can be partly developed according to its own regularities. We consider internalization as a process of mutual regulation of external and internal activities.

6 CONCLUSION

The analysis of Rubinshtein, Vygotsky and Leont'ev works shows that each emphasized different theoretical studies and different aspects of activity and individuality. Each of their contributions into AT is important. Both Rubinshtein and Leont'ev proceeded from the general statement that object-practical activity is the principal source of mental development. Vygotsky was primarily studied personality from the cultural-historical perspective. Internalization for him was first a social process, in which the semiotic mechanisms, particularly language, mediate the interaction between social and individual functioning. Individual practical aspects of activity were underestimated in his theory. Rubinshtein did not accept the idea of internalization. However, he did not consider the semiotic aspects of activity sufficiently, nor did he discuss the mechanism of self-regulation, which explains the interdependence of the external and internal activities more precisely. His major point was, "external causes act through internal conditions". Leont'ev did not give sufficient value to the role of internal conditions in development.

In SSAT, the interrelationship between the external and internal components of activity during the mental development is considered from the perspective of self-regulation. Human behavior cannot be reduced to the external (stimulus - response) manifestation of activity. In AT, the person who interacts with a situation is considered the subject. That is, we talk about action and cognition, not about the stimuli to which the subject reacts. Personality is developed through a person's participation in activity, which depends on the relationship between the subject and situation and the relationship between subjects. That is, activity is both external and internal.

REFERENCES

Bedny, G. Z. and W. Karwowski. 2007. *A Systemic-Structural Theory of Activity. Application to Human Performance and Work Design*. Boca Raton, FL: Taylor & Francis.

Bedny, G., Karwowski, W. and Voskoboynikov, F. 2010. The Relationship between External and Internal Aspects in Activity Theory and Its Importance in the Study of Human Work. Bedny, G. and Karwowski, W. Eds. *Human-Computer Interaction and Operators' Performance*. Boca Raton, FL: Taylor & Francis.

Leont'ev, A. N. 1978. *Activity, Consciousness and Personality*. Englewood Cliffs, NJ: Prentice Hall.

Rubinshtein, S. L. 1940. *Problems of General Psychology*. Moscow: Academic Science.

Skinner, B. F. 1974. *About Behaviorism*. New York: Knopf.

Vygotsky, L. S. 1962. *Thought and Language*. Cambridge, MA: MIT Press.

CHAPTER 13

Positioning Actions' Regulation Strategies

G. Bedny, W. Karwowski
Evolute
Louisville, KY, USA
gbedny@optonline.com
University of Central Florida,
FL, USA
wkar@ucf.edu

ABSTRACT

This study examines motor positioning actions performance from systemic-structural activity theory perspectives. Assembly-line workers, operators and computer users in human-computer interaction and man-machine systems often use these types of actions. In this study we demonstrate that Fitts' law doesn't adequately describe real strategies of human performance when positioning motor actions are utilized.

Keywords: self-regulation; strategies of performance, pace of performance, precision; errors analysis.

1   INTRODUCTION

In manual tasks there is often a need to move small objects into a particular position. To complete such tasks, a worker performs various positioning motor actions. For example, an assembly-line worker needs to move parts into a certain position. In man-machine or human-computer interaction systems there is a need to move a joystick to a particular position, or put a cursor at a certain point on a screen. Such actions do not require physical efforts. The main requirements for these actions are time limit and precision of movement. Fitts (1954) investigated the relationship between three variables: time, accuracy (precision), and distance. In his study subjects had to move their hand from one target to another, and hit them with maximum speed. Fitts' studies have shown that movement time is linearly related to the logarithm of the *index of difficulty*. His studies were summarized by an equation presently known as Fitts' Law. Some authors suggest using Fitts' Law to predict positioning actions performance time. When attempting to apply this Law to a work environment, one would assume that each operator's action is independent, and each operator's action is performed with maximum speed. It is difficult to agree with such assumptions. Actions during task performance are organized into a system, and influence each other. A subject does not react, but rather actively and purposefully acts during task performance. Pace of performance of such actions is much lower than stated in Fitts' Law.

In this work we describe performance strategies for positioning motor actions. Unlike previous research that studied positioning motor actions with two targets, this research considers not just two, but four targets as well. We've obtained new data related to the properties of the regulation process for positioning actions and discovered new opportunities for application of these properties. Positioning motor actions are observed from the systemic-structural activity theory (SSAT) point of view. In this study functional analysis of positioning actions has been conducted while activity is considered as a self-regulative system. The study of action precision and error analysis in this work is performed from the activity self-regulation viewpoint. It is also demonstrated that not only cognitive but also emotionally-motivational mechanisms of activity regulation are important in errors analysis. This aspect of error analysis can't be reduced to studying performance under stress. We demonstrate that data and methods developed in the framework of SSAT can be useful in studying positioning actions regulation and in predicting their performance time.

One important aspect of performance time is the pace or tempo at which a person is working. Therefore, when studying positioning motor actions, pace of performance should be taken into consideration. Another important aspect in studying positioning motor actions is that they are almost never performed in isolation from other actions. These other actions are logically organized, and influence performance time of positioning motor actions. However, in Fitts' experiment, the focus was on how the subject performed the same positioning motor action multiple times with maximum speed. The purpose of our study was to analyze whether there were changes in the performance time of positioning motor actions when there are two and four targets.

2 CONCEPT OF SELF-REGULATION IN SSAT

There are two main methods in SSAT for studying activity: morphological and functional. In the morphological study, action and operations are the main units of analysis (Bedny, Karwowski, 2007). One can distinguish between external behavioral and internal mental actions. External behavioral actions include various motions and transformation of material or tangible objects. Mental actions transform images, concepts or propositions and non-verbal signs in mind. Functional analysis is based on studying mechanism of activity self-regulation. The concept of self-regulation is critical for understanding activity as a system and for analysis of preferable strategies of performance. In activity theory, self-regulation is not a homeostatic, but rather a goal-directed process where the goal has an integrative systemic function.

The concept of self-regulation is more meaningful when the self-regulation model is developed. This model is defined in terms of functional mechanisms or function blocks. At any point of time, cognitive processes are integrated to achieve a specific purpose of activity self-regulation. This integration of cognitive processes is the basis for the formation of function blocks or mechanisms of self-regulation. A number of function blocks in a self-regulation model are constant. However, the context of these blocks changes constantly. The SSAT concept of self-regulation is different from the concept of self-regulation outside of activity theory where models of self-regulation are homeostatic. Homeostatic models function as discrepancy reduction systems that can't prevent errors. However, people often intentionally make errors to evaluate consequences. The goal in homeostatic models of self-regulation is viewed as an unchangeable readymade standard. In reality, the goal is interpreted, accepted, or reformulated by a subject. During task performance, the goal becomes more specific and clear. In cognitive psychology goal and motives are not clearly distinguished. For example, in cognitive psychology goal has such characteristics as intensity. SSAT distinguishes goal from motives. Motive and goal create vector motives→ goal. Motives are energetic and goal is cognitive components of activity. This vector gives activity goal directed character. There is only one goal of task and usually several motives for its performance. In contrast to cognitive psychology where goal can be conscious or unconscious in activity theory goal always includes conscious components.

In SSAT there are general model of self-regulation and model of self-regulation of orienting activity. Here we consider self-regulation model of orienting activity to describe how a person mentally reflects and interprets a situation during performance of positioning motor actions. This stage precedes task execution. From this point of view orienting activity is to some degree similar to the concept of situation awareness in cognitive psychology. However, orienting activity is a much broader concept that includes goal, conscious and unconscious components, subjective criteria of success, etc. Stable and dynamic mental models can't be fully verbally and described by a subject. Explorative actions and operations play an important role in mental representation of reality.

Self-regulation model of orienting activity includes only 13 blocks (see figure 1). During task analysis, the researcher usually takes into consideration not all function blocks, but only those that play the most important role in the considered task performance. This simplifies usage of this method of study. The researcher should pay attention not only to separate function blocks, but also their interrelationships. A model of self-regulation of activity can be interpreted as an interdependent system of windows (function blocks) from which we can observe human activity during task performance. For example, a researcher can open a window called "goal" and at this stage of analysis can direct his attention to aspects of activity such as goal interpretation, goal reformulation or formation, goal acceptance, the relationship between verbally logical and imaginative aspects of goal, possibility to understand the goal at different stages of task performance in more detail, influence of goal on interpretation of input information, and so on. In contrast, when we open the block (window) "assessment of the sense of input information" we concentrate our attention on such aspects of activity as significance of input information and goal, etc. Hence, this block is associated with emotionally evaluative aspects of activity.

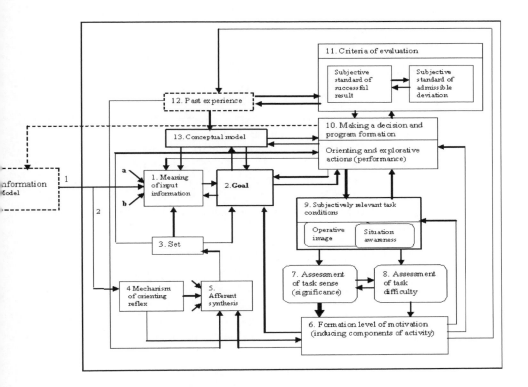

Figure 1. The model of the self-regulation of orienting activity.

Let us consider this model in a more detailed manner. The model includes both conscious and unconscious levels of self-regulation, which interact with each other. In our discussions we consider an example of unconscious level of self-regulation. The dashed box "informational model" is not a function block. It is the information that is presented to the performer. Let us consider the unconscious level of self-regulation (channel 2). The incoming, and usually unexpected, information activates the orienting reflex (block 4). The orienting reflex is conveyed by responses such as turning the eyes or head toward the stimulus, altering the sensitivity of different sense organs, changing the blood pressure and heart rate, etc. At the same time, there appears to be some electrophysiological change in the activity of the brain. The orienting reflex plays an important role in the functioning mechanisms of involuntary attention. The orienting reflex provides automated tuning to external stimuli and influence on the general activation and motivation of the subject (see the connection between blocks 4 and 6). The orienting reflex also influences afferent synthesis (see interconnection between blocks 4 and 5). The horizontal arrow demonstrates major or relevant information that initiates a response or orienting reflex. However, the major stimulus never exists in isolation. The environmental background is also a source of additional (situational) information. These situational stimuli have some influence on the response to the major stimulus. Hence, the afferent synthesis mechanism (see block 5) can also receive irrelevant stimuli (diagonal arrows). For example, non-instrumental information or irrelevant environmental information can have an influence on afferent synthesis in combination.

Afferent synthesis (block 5) is also affected by the block of motivation (see connection between blocks 6 and 5) and past experience (block 12). Therefore, afferent synthesis (block 5) performs an integrative function by integrating temporal needs and motivation, mechanisms of memory representing relevant past experience, and the effect of both irrelevant (non-instrumental) and relevant (instrumental) information. Afferent synthesis helps a subject to select with some degree of accuracy an adequate major stimulus from among the infinitely varying influences.

Influences from afferent synthesis promote the formation of block 3 (goal-directed set). This set is formed by instructions and the specifics of the situation. Goal-directed set should be distinguished from a relatively stable system of sets, which is a result of life experience. Goal-directed set is close to the concept of goal but not sufficiently conscious or completely unconscious. Such a set gives a constant and goal-directed character to the unconscious components of activity. It helps to retain its goal-directed tendency under constantly changing conditions. Goal-directed set connected with block 1 (meaningful interpretation of information). Due to the lack of sufficient activation of block 2 (image-goal), which is a major integrative mechanism of conscious regulation of activity, meaning block 1 is primarily non-verbalized. Hence, unconscious processes and motivation affect our cognitive mechanisms. If activity regulation is predominantly unconscious, then a "goal-directed set" has a direct impact on block 10 (making a decision about the situation or strategy of explorative actions). If the "set" is inadequate, farther interpretation of the situation also can be incorrect. As a result, explorative actions associated with block 10 cannot have a purposeful or goal-directed character and become chaotic. During unconscious information processing (channel 2) the goal (block 2) is not activated. Connection between block 3 (goal;-directed set) and block 2 (image-goal), where goal is conscious demonstrates that "set" in some situations can be transformed into conscious goal. Block 2 and associated with it other blocks reflect the conscious goal directed regulation of activity.

The following blocks are especially important in analyzing task performance strategies for the task under consideration: goal (block 2); subjectively relevant task conditions responsible for development of stable or dynamic mental models of the task or situation (block 9); assessment of task difficulty (block 8); assessment of sense of task or task significance (block 7); formation of level of motivation (block 6); evaluation criteria (block 11) includes two sub-blocks: "subjective standard of successful result" and "subjective standard of admissible deviation."

3 METHOD

From the SSAT perspective, it is important to study structure of activity while a subject uses various strategies of task performance. Hence, one should utilize experiments where conditions of the tasks performed vary by including or excluding elements of activity, changing sequence of their performance, etc. to discover their interrelation. Objective measurement procedures are combined with such subjective methods as observation, verbal discussion, and error analysis. In our experiment we introduced two main conditions of task performance: 1) subjects performed positioning actions with two targets; 2) subjects performed positioning actions with four targets. Distance and width of targets also varied. Figure 2 shows the layout of four targets.

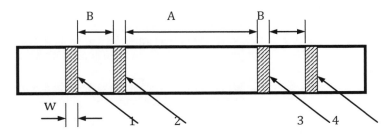

Figure 2. Layout of four targets. 1-4 –targets; A and B – distance between the targets; w – width of the targets.

In one series of experiments the subject was instructed to strike two targets with a metal stick, while in the other series of experiments, four targets were involved. The purpose was to understand the strategies of positioning motor actions regulation, when such actions are performed as isolated actions and then in sequence.

As soon as the red bulb was turned on, the subject raised and held a metal stick (connected electrically to the meter) above the target without touching the apparatus (start position). Two seconds later, the green bulb was turned on, two meters were turned on and a stopwatch was started. The subject was trying to strike targets at his/her maximum speed. After ten seconds, the buzzer signal was activated and the apparatus was turned off. By tracking the entire task performance time and the number of times the subject successfully struck a target the average action performance time could be determined.

The experimental group consisted of five university students. They were instructed to hit the target with maximum speed and precision. (Fitts gave the same instruction to his subjects). Two groups of experiments were conducted. In the first group of experiments, only wide targets were used (w = 50 mm). In the second group of experiments, only narrow targets were used (w = 7mm).

The main purposes of this study were to discover the characteristics of positioning motor actions conducted in isolation and compare them with the sequential performance.

4 RESULTS AND DISCUSSION

The first two function blocks that essentially influenced a strategy of positioning actions performance were the block "goal" (block 2) and "subjectively relevant task conditions" (block 9). Together these two function blocks are responsible for creation of a mental model of task or mental representation of task. Findings supported that the instruction "hit the target at your maximum speed and precision" gave the subject an opportunity to vary widely his/her subjective mental representation of the task. Some subjects considered precision to be the main requirement of the task while others considered speed of performance to be more important. Relationships between these two requirements vary substantially among subjects and tasks. As a result, strategies of task performance also vary. In some cases, it was necessary to introduce additional corrective instructions during the training sessions until the subjects adapted to experimental requirements.

Let us consider some experimental data.

In the first series of experiment subjects hit only two wide targets. Function block 8 *"assessment of task difficulty"* provides an explanation of how subjects achieve required precision of performance when they strike two wide targets in the first group of experiments. Tapping 50-mm wide targets is seen by subjects as a relatively easy task, because these targets are subjectively wide enough to comply with precision requirements. These requirements do not contradict with the time demands, and the subjects feel they can manage both precision and speed requirements equally. This feeling influences motivational block 6 and emotionally-evaluative block 7. The subject is motivated to follow instructions. Significance of speed requirements increases along with precision significance. The subject becomes motivated to avoid errors. This, in turn, influences such mechanisms of self-regulation as subjective standard of successful result and subjective standard of admissible deviation from the standard (the first and the second mechanisms of block 11).

It was shown that during the training process and the period of the experiment, the subjective standard of success was formed. This subjective standard does not always match the objective standard of success (objective width of the target). Approximately 80% of the hits were placed in the middle of the target with 35 mm range. This means that the subjective standard of success was much narrower than the objectively given standard.

In series 3, subjects struck four wide targets. In this series, the subjects had a tendency to alter their criteria for success in comparison to task when subject hits two wide targets. . The target zone was expanded but remained narrower than the entire width of the targets. The choice of a subjective standard was determined by the width of targets, distance between targets, individual criteria of success, and personal ability of a subject to achieve a set goal "to hit targets with maximum speed." The functional mechanism *subjective standard of success* is tightly connected with the concept of "reserve of precision," which both has objective and subjective meanings (Bedny and Meister 1997).

Under reserve of precision, we understand the difference between the width of the target and subjectively selected target area that the subject is trying to hit during trials. The greater the width of subjective standard of success (width of hitting area of target), the smaller the subjectively acceptable reserve of precision, and the subjects employ an increasingly risky strategy.

This is manifested by the fact that the hitting area in the target becomes broader and approaches the edges of the targets. Hence, the subjects begin to use a more risky strategy, because they approach the edges or borders of the targets. However, the subjects still avoided approaching the edges of the targets. This strategy allows subjects to maintain the same objective criteria of success (*do not hit areas outside of the targets and sustain required speed*). Such a strategy helps to reach the objectively presented goal "hit targets with maximum speed." Qualitative analysis of activity strategies is confirmed by error analysis.

Let us consider some subjects' strategies when they hit narrow targets. The functional mechanism of block 11, *subjective standard of successful result,* and its relationship with other blocks when subjects use narrow targets becomes particularly important in error analysis. In the case of the narrow targets, subjects used the full size of the target when selecting their subjective standard of success. The subjectively acceptable reserve of precision is close to zero. Small deviations from a given mode of actions could result in missing the target (risky strategy). Most subjects consider tapping narrow targets a much more difficult task than tapping wide targets. Attempts to use a narrower area inside the target might significantly increase performance time. At the same time, however, an increased number of errors for difficult task performance are considered more acceptable than sharp decreases in speed of performance. In experiments involving four narrow targets, the errors were considered critical by most of the subjects when they missed several targets during one sequence of movements. When a subject missed just one of the targets with minor deviation, she/he started correcting such errors after only several occurrences. Hence, a mechanism such as *subjective standard of admissible deviation* (block 11) also plays a significant role in self-regulation of positioning actions.

This experiment demonstrated that subjects alter their strategies of attention through the series of experiments. When a subject worked with two targets, she/he tried to hold them in her/his field of view (divided attention strategy). However, in case of four targets, one not only needed to distribute attention, but also to switch it appropriately.

The subjects combined targets into viewing groups based on the distance between the targets. Consequently, the first and third actions with smaller inter-target distance were considered primary, and the second action with the large inter-target distance was viewed as auxiliary, serving only as the means of going from the first to the third action (Figure 2).

The presence of rigid, external instructions within a planned experimental environment often does not uniquely determine subject's activity. A subject creates her/his own representation of situation, and develops and evaluates strategies based on personal capabilities and activity demands. This often complicates transfer of laboratory results to field application, especially in case of studying isolated actions, movements, and psychological processes. The qualitative and quantitative analysis of strategies used by the subjects indicated various modes of self-control and corrections of activity, which in combination serve as a self-regulative process.

This research indicates that a person is not merely reacting to external factors (stimuli). Subjects actively interact with their environment. They select their own goals, reformulate tasks, and change their subjective standards of success and strategies of activity. In order to be able to apply experimental results, the experiment should be targeted to explore typical strategies of performance in a dynamic environment.

Our studies have shown that positioning motor actions can't be seen as independent in the context of the entire activity. This leads to the conclusion that Fitts' law can only be used to measure isolated, discrete, motor actions performed with maximum speed. In addition, our studies demonstrated that when one switches from discrete positioning actions (two targets) to sequential positioning actions (four targets), there is a significant increase in action time for low-precision actions (wide targets). Here we found a change from an automatic level of self-regulation (in case of two wide targets) to a conscious level of self-regulations when subjects hit four wide targets. Switching from two to four narrow targets causes the change toward a higher level of self-control within the conscious level of self-regulation. This explains why the pace slows down less when subjects transfer from two to four narrow targets than during the transfer from two to four wide targets.

Further, it was noted that distance between targets influences the high precision movements, but it does not significantly impact low-precision movements. These results should be taken into consideration when designing efficient manufacturing work processes and equipment. Study of precision in human performance demonstrates that it can't be reduced to just quantitative aspects of error analysis. The inadequate strategies of performance causing the errors should also be considered. Discovering such strategies and studying their causes is very important in improving precision of performance and reducing errors. Emotional motivational and cognitive aspects of activity are very important for improving precision of performance. The change of errors' significance for subjects (emotionally-evaluative aspects of error analysis) and motivational state of subjects can dramatically change the precision of their performance. This aspect of studying precision of task performance can't be reduced to traditional error analysis of operator performance in stressful conditions. Moreover, in stressful situations subjects use a variety of strategies to adapt to emerging conditions.

REFERENCES

Bedny, G. Z. and Karwowski, W., 2007. A systemic-structural theory of activity.
    Application to human performance and work design. Boca Raton: Taylor and Francis.
Fitts, P.M., 1954. The information capacity of the human motor system in controlling the
    amplitude of movement. Journal of Experimental Psychology, 47, 381-391.
Ponomarenko, V. and Bedny, G. (2011). Characteristics of pilots' activity in emergency
    situation resulting from technical failure. In G. Z. Bedny, and W. Karwowski
    (Eds.).Human-computer interaction and operators' performance. Optimizing work
    design with activity theory. Boca Raton: Taylor and Francis, 255-276.
Bedny G. Z. and Meister, D. 1997. The Russian theory of activity. Current applications to
    design and learning. Mahwah, New Jersey: Lawrence Erlbaum Associates, Publishers.

CHAPTER 14

Emotional-Motivational Aspects of a Browsing Task

von Brevern, Hansjörg; Karwowski, Waldemar
Independent Consultant; University of Central Florida
Zürich, Switzerland; Orlando, USA
vonbrevern@acm.org; wkahfe@gmail.com

ABSTRACT

Motives and their congenital goals are individual, vary with every human subject (HS), and coexist in space and time. They are dynamic and as such formulate the contingent nature of the social aspect of activity. Overall, subjective goals, motivation, and significance may contradict with objective stimuli. This paper presents a case study of a HS who skims through news feeds on the corporate intranet to explore possibly relevant information. One subjectively significant news feed causes him to temporarily interrupt work activity and instead follow up a link where he can make a decision regarding his next year's contribution of the local pension fund with its 13'236 members. Contrary to his intention to instantly attain his goal, his task strategy succeeds in undesirable abandoned actions, which gradually increase frustration. Under the umbrella of the applied systemic-structural theory of activity (SSAT), we discuss emotional-motivational effects on task performance, manifest differences between the holism of human goal-oriented activity and engineering, contrast objective design goals with subjective goals of activity, and present issues and findings from informal and semi-formal qualitative analysis (QA) of our HS's exploratory behavior and rational decision-making. Theoretical and pragmatic discoveries encourage us to postulate an ameliorated task strategy. We finalize this paper with possible future research prospects.

Keywords: Systemic-Structural Activity Theory (SSAT), motivation, qualitative analysis (QA), error analysis, decision point (DP), abandoned actions, task strategy

1   INFORMAL QUALITATIVE ANALYSIS

Deficiencies of personal assistance, individualization, customization, dynamic guidance, and adaptation (von Brevern and Synytsya, In press) are not only an issue of complex systems within large corporations but are an everyday experience – be it on the Intra- or Internet. While work activity is goal-oriented, browsing is a purposeful exploratory activity that usually consists of simultaneous-perceptual actions, which are inherently mental. Simultaneous-perceptual actions identify, perceive, and logically and mentally distinguish between stimuli, events, and objects. So, have we never wondered why some hypermedia sites on cyberspace are engaging while others trap us in a maze of frustration until we capitulate? At first sight, both may look appealing with harmonious color schemes, tasteful graphics, and embedded menus from usability views but their responses to our subjective behavior may ultimately substantially affect our motivational-emotional states.

Unlike empirical analyses that essentially abstract results from a multitude of HS, our study and discussion intentionally entails the behavior of only one HS who decides to temporarily suspend his ongoing work activity to skim through the news feeds on the corporate intranet portal on the search for 'something interesting'.

Just as our HS's eyes are skimming through the three news sections of the 'home' page from top to bottom, one heading at the third and last section suddenly attracts his full attention: "Pension Fund: Choose Your Contribution Option in the Savings Plan". This perceptual stimulus immediately changes his focus of attention. In previous years, an email alerted internal staff to select one's personal contribution of the pension fund for the coming year, it is just one amongst many other inconspicuous news feeds this year. This unexpected discovery perplexes and yet excites him. The pension scheme is relevant at any age for retirement. So, he decides to immediately substitute all objectively given work goals and tasks and presumes that making his choice will be done quickly.

Individual significance closely affects the interaction between cognitive and emotional-evaluative mechanisms that mutually stimulate a HS to select and interpret information presented at the computer screen. This phenomenon influences the cognitive strategy of activity when HS interact with computers and might explain why HS accept and pursue inadequate goals of activity for a specific task. Our HS' activity is explorative because of unknown sequences of actions in advance, correctness of the strategy of task performance, attention, and time required. Consequently, an erroneous strategy may result in abandoned cognitive and motor actions leading to an undesirable result. So, the strategy of task performance ensues his subjective mental decision to choose next year's contribution for the pension fund. Hence, his motives and goals are subjective.

On the 'home' page, our HS then focuses on the link of the news feeds, moves the cursor to the link, presses the mouse button, and releases it. These actions are members of algorithm, which are actions "integrated by a higher hierarchical goal into some combination of actions" (Bedny and Karwowski, 2007, p. 79). Their higher hierarchical goal is to 'open the newsfeed'.

The browser then takes our HS to the news feed. He only glances through the long and scrollable text hoping to find the very link that will bring him smoothly to the submission page. Unfortunately, his hope is about to be dashed. The bold typeface of the first paragraph of the news feed attempts to grasp readers' attention but requires concentration and mental reflection for it presents two preconditions of which one involves extraction of information from long-term memory: ((born >= 1971) XOR (joined pension plan > 1.1.2010)). The second paragraph in regular typeface describes how to navigate to the submission page and informs about the last date of submission: "Go to MyHR+ and "Salary, Compensation & Benefits" no later than December 15, 2011 to choose the amount of your savings contribution. You can choose one of the three alternatives: Basic, Standard, or Top". This paragraph does not contain any link to the submission page. The phrase in blue font color "→ Choosing the contribution option in the savings plan" beneath the second paragraph links to the contribution plan yet seems unrelated with the preceding context. Finally, the last section exhaustively describes each type of the available contribution levels in detail. In all, there is no recognizable button or link that at least guides visitors directly to the submission page. A reader in a rush will tend to scroll to the bottom of the page to look for a corresponding link or button.

Therefore, our HS decides to navigate to the submission page manually as described in the second paragraph. Doing so requires him to keep instructions in working memory, which is undesirable for any task and to be avoided at the design stage. As result, our HS starts a comprehensive and extremely time-intensive journey with searches, clicks, and navigation through numerous pages like "MyHR+", "Salary & Compensation Benefits", "Payroll & Compensation Benefits", "Contribution Levels under Savings Plan", "OneHRIS", the search center, and in-page searches. Although, explorative behavior using hypertext-like applications facilitates the correction of 'reversible errors', his goal-directed behavior is gradually becoming chaotic as exploration continues. The evidently simple task to choose and save his personal contribution has become a frustrating marathon through the corporate cyberspace. With each return to the news feed, he only focuses on the second paragraph and never notices the first paragraph and the link beneath the second paragraph. However, returning to the news feed is useless and time wasting because instructions are incomplete, incorrect, confusing, annoying, and frustrating. After a laborious journey, our HS finally reaches the submission page where he makes his choice and presses the save button. Without any feedback from the system, he resumes his work and expects his account to reflect his change early next year. Personal significance has stimulated his motive that has driven the goal to not abort his task. Ultimately, our HS feels pride about his goal-achievement and satisfied to have 'conquered' the unsocial system.

Per contra, our HS will never receive his statement of account with the expected contribution. Without him either realizing or being advised by the system, he does not fulfill any of the pre-conditions. So, our HS has become a victim of 'unsocialization' by the system where the "causes of behavior are often to be found in the situation rather than the person, and sometimes it takes a kind of Sherlock Holmesian perspicacity ... to detect such causes" (Abelson et al., 2004, p. 218). Howbeit, the designer of the system has not had a Holmesian eye and may blame HS as being liable for their behavior. Our case demonstrates the criticality and significance of the subjective motive, which could be one plausible reason why HS tolerate unsocial system behaviors.

2 HUMAN ACTIVITY CHALLENGES COMPUTING DESIGN

Conventional software engineering (SE) and human-computer interaction (HCI) bestow a 'stimulus-response' interaction between the system and its environment. On the basis of our case, the subject domain of the first interaction in SE is opening and displaying the news feed. The major concern of HCI and human factors is the assessment of the design of products and the use of systems so that the first interaction would look at the design and use of software and hardware artifacts (e.g., eye movements, hand movements, and the ease of use of the mouse). Their view is thereby narrower than the one of SE while both seek to elaborate the behavior of artifacts interpreted and modeled on actual human utilization. Their shortcoming is the understanding and prediction of human action that precedes any stimulus-response man-machine interaction. SE and HCI only have a vague understanding of human activity, task, and actions and largely ignore the motivational-energizing effects of human motives on the goal of activity. The root of our HS's behavior is his motive with its emotional-motivational significance to select his contribution for the following year's pension scheme. This motive energizes his goal and ultimately prevents him from giving up despite his insidiously increasing frustration. Conventional engineering is unable to recognize granular human mental behavior on tools or artifacts that precedes the known stimuli-response behavior on man-machines. As a result, anticipated savings on the design by the organization may be ultimately exceeded by labor costs and efforts spent on searching, navigating, and the handling of latent claims. A discussion on stimuli onto the design from its organizational environment is outside of the boundaries of this paper but has been adduced by (von Brevern and Synytsya, In press).

3 COMPREHENDING HUMAN ACTIVITY WITH SSAT

SSAT proffers an atomic understanding of human goal-oriented activity and applies four stages of analysis i.e. QA, algorithmic analysis (AA), time structure description, and quantitative analysis. SSAT views activity as "a goal-directed system, in which cognition, behavior, and motivation are integrated and organized by a mechanism of self-regulation toward achieving a conscious goal" (Bedny and Karwowski, 2007, p. 1). The vector 'situational and sense-formative motives→ goal' (Bedny et al., 2012) "lends activity directness and is the energetic component, which drives activity towards the achievement of a specific goal" (Bedny and Karwowski, 2007, p. 314). The basic object of study is the task, which "involves goal achievements that require motivational forces", "always include[s] more or less problem-solving aspect[s]", and "is a situation which requires achievement of a goal in specific conditions" (Bedny and Karwowski, 2007, p. 37). Actions are the discrete elements of an activity "that fulfills an intermediate, conscious goal of activity" (Bedny and Karwowski, 2007, p. 41) and are responsible for achieving the goal of the task during subject-object mediated activity on physical or mental artifacts or tools. The concept of human actions as recognized by the Russian AT has largely affected the development of "Action Theory" in Western Europe. However, unlike action theory and cognitive approaches, actions in AT can e.g., be decomposed into cognitive and behavioral components, contain goals, be interlinked with motivation, be organized in a linear sequence, contain mechanisms of information processing, be a tool for the creation of images, and are therefore the base for precise systemic and structural analysis. From activity-theoretical perspectives, the "end of an action can never serve as the beginning of another action" (Anokhin, 1966, p. 77). This perspective thus radically differs from the plain stimulus-response interaction in SE and HCI. In SSAT, each single action is still a discreet element of human goal-oriented activity with its self-regulative feed-forward and –back loops consisting of input, goal acceptance/formation, goal evaluation, decision-making/execution, evaluation of result, and correction. Consequently, the study of action can have multiple foci on e.g., a probabilistic prognosis, goal-related, emotional-motivational, spatial, or temporal phenomena.

As per the hierarchical structure of human activity, SSAT recognizes all of its four stages as hierarchically organized methods of analysis permitting iterative cycles. This enables decomposing activity at different stages into varying degrees of depths. QA as antecedent of AA portrays an objectively logical method to study task performance. Per se, this stage is not restricted by any specific formal method but requires a delineation of e.g., job performance, tasks, technological (sub) processes, artifacts, the relationship of technical equipment, work, and environmental, social, and cultural conditions. However, QA acknowledges various numbers of methods. AA is a morphological description of activity and consists of "qualitatively distinct psychological units and determination of the logic of their organization and sequence" (Bedny and Karwowski, 2007, p. 87). Mental and motor actions are thus central. They are either 'operators' or 'logical conditions', are embedded into a logical and sequential structure, interrelate, and can generate other members. Closely interdependent homogeneous human actions are bound in time and space by their sub- and superordinate goals. The goal of activity unites them into a holistic system, which shapes a hierarchically organized, logical, and meaningful structure (Bedny and Karwowski, 2007). More formally, this structure contrives a human algorithm of task performance in situ. Ergo, SSAT is suitable for analyzing new systems, for anatomizing the structure of activity, for task-goal decomposition, for human reliability assessment, for reengineering, for error analysis, and for improving task performance.

4 SEMI- & FORMAL QUALITATIVE ANALYSIS

Despite the HS's high task significance, numerous factors elevate our HS's inefficient strategies of task performance. For example, we do not know design rationales for overly repeating conditions, stressing the capacity of short-term memory (e.g., 7 ± 2 objects), neglecting that HS do not analyze all available information at any given point of time and attention-switching that may affect task significance or even lead to reformulate the goal of activity, inconsistent naming and verbal descriptions, lack of external feedback, and absence of system validations. Our case study demonstrates that our HS's perception differs from the designer's intention – resembling a leap of faith. More precisely, objectively defined goals do not coincide with subjectively formulated and accepted goals. Wasted labor, time, and emotional costs could have been avoided, if the design had incorporated actual "thinking actions that usually precede decision-making actions. Such actions include categorizing objects by certain criteria, making inferences, taking actions that reveal the functional relationship between the elements of the situation, and so on" (Bedny and Bedny, 2010, p. 159). Following our informal inquiry into task performance, we have applied the semi-formal 'Business Process Model and Notation' (BPMN) because of its shared understanding between business, analyst, and developers. Akin to human behavior, BPMN allows hierarchical decompositions of processes, sub processes, tasks, actions, pools, and lanes. Unlike SSAT, BPMN bears a mixed understanding of activity because it seeks to satisfy human, technological, and process-oriented interactions: BPMN's view of *activity* can be atomic, compound, a *process*, or even, for example a user, business, or script *task*. Figure 1 illustrates our semi-formal decomposition of the navigational behavior on the 'home' page and models 'top-down' browsing behaviors in probabilistic manners. These behaviors correspond with subjective motivational significance after the decision points (DP) d_0, d_1, d_2. Probabilities of actions that proceed from d_1 are $(p_1 + p_2 + p_3 + p_4 = 1)$ resp. $(p_5 + p_6 + p_7 + p_8 = 1)$ from d_2. At this stage, task significance of the goal of activity from d_0 and the probabilities of the strategies of work performance (i.e., $(p_3 = 0.2)$, $(p_7 = 0.2)$) that lead to the 'community news' are low. However, our HS' goal of activity and its state of significance immediately change to 'high' when he discovers the link "Pension Fund: Choose Your Contribution Option in the Savings Plan" in the 'community news' despite the originally low probability $(p_3 = 0.2)$ after d_1 of the previous goal of activity. Meanwhile, our HS has spent little time from d_0 to the 'community news' $(d_1, (p_3 = 0.2))$ with a low motivational significance. His meaningful discovery results from direct connection actions that motivate decision-making.

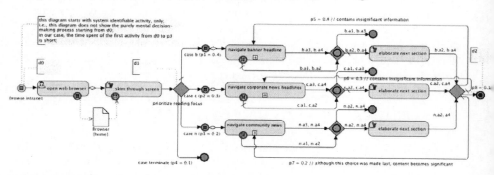

Figure 1 Predictive navigational behavior of the HS through the news feeds on the 'home' page

Probabilities of his task strategy reflect subjective significance, which is an expression of decision-making embracing mental transformational and/or direct connection actions. These actions are part of self-regulative human activity, which emphasizes the criticality of decision-making in goal-acceptance, execution, and the evaluation of a result (Synytsya and von Brevern, 2010, p. 401). Altogether, a DP formulates design challenges to reflect successive actions. From the angle of a DP, successive actions are predictive and driven by subjective significance. A basic prediction from the eye of a DP is the reflection of actual human behavior in engineering that stems from human observation with an existing system; a more challenging type of prediction of human behavior is one where a system does not yet exist. Hereafter, we focus on basic prediction for error analysis and deduce an ameliorated solution. SSAT distinguishes between deterministic (i.e., 1, 0) and complex (e.g., AND, OR, and IF-THEN) DP. BPMN allows 6 types of DP i.e., the unspecified, parallel, exclusive, inclusive, event-based, and complex DP. While human decision-making is purely mental, the subject domain of a system in SE (and inherently in BPMN) considers only identifiable stimuli. Although the shallow argument prevails that design should only select the most typical task strategies, this assertion neglects the fact that an accurate selection demands for a universe of all possible task strategies.

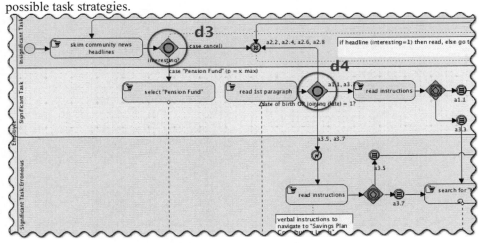

Figure 2 Low-level extract of the navigational behavior of the HS to opt for the pension fund

The computation of the universe X requires identifying all possible logical combinations of conditions c so that $X = 2^{\sum C_1, C_2, ..., C_n}$. X allows us to construct a decision table that is a matrix between c and all combinations of c, to decide whether true (x) or false () for each set of conditional combinations, and to finally assign predictive actions for each plausible set. Moreover, one plausible set may have multiple probable task strategies. The subject domain of d_4 in Figure 2 consists of the 2 basic and identifiable system conditions i.e., '((dateOfBirth >= 1971) OR (joinedRetirementsPlan > 20100101)) AND (submissionDate =<20111215)' and selection of the link "↪ Choosing the contribution option in the savings plan" so that $X_1 = 4$. The system can identify these 2 stimuli because it can detect and retrieve employees' data in the background. Based on X_1, the decision table A of Figure 3 illustrates that all sets are plausible but we can neither assign our HS' actual behavior to any set nor detect error scenarios. Consequently, we have enriched d_4 in Figure 2 with d_3, which is the addition of our HS's mental decision to follow up the link on the community news section with its headline "Pension Fund: Choose Your Contribution Option in the Savings Plan". As discussed in chapter 1, our HS' decision-making and motor actions are members of the algorithm of the sub-goal of the sub-task to 'open the newsfeed', which has stimulated and motivated the significance of the new goal of activity. In view of the subject domain, the system can identify the HS's motor action. As a result of our enhancement of d_4 so that $X_2 = 8$ in B of Figure 3, we are now able to assign all multiple strategies to each plausible set, detect the erroneous and frustrating maze of our HS's task execution (c.f., matrices of a3.3, a3.5, and a3.7 in Figure 2 and B of Figure 3), and can catch the designer's misalignment with our HS's cognitive perception (c.f., 'X/X1' of the matrix a3.3 in B of Figure 3). As a result from X_2 and in view of the solution space, these 3 conditions must be integrated into only 1 man-machine interaction, which the computer should validate, and if eligible, take the HS directly to the submission page or inform about our HS' non-eligibility.

A: Full set of conditions: Stimulus-response conditions only

$\# Combinations = 2^{\sum C_1 C_2 \ldots C_n} = 2^2 = 4$

Title Does the employee satisfy the conditions and discover the link beneath the instructions paragraph? (without mental action)

#	Conditions (C)	# Combinations 1	2	3	4
C1	((employee.date of birth >= 1971) OR (employee.joined retirements plan > 20100101)) AND (submission =<20111215)	Y	Y	N	N
C2	the employee sees the link "→ Choosing the contribution option in the savings plan", assumes that the link will lead to task accomplishment	Y	N	Y	N
	Deductive scenarios:				
	Reasonably possible scenarios	X	X	X	X
	Error scenario				
	Actions				
A1	read instructions; discover link "→ Choosing the contribution option in the savings plan"; clicks link "→ Choosing the contribution option in the savings plan";	X			
A2	cancel (employee does not want to change contribution options); reads OR does not read the rest of the page;	X	X	X	X
A3	which action(s) will the employee do?		X	X	X

= reasonably possible scenario by combination

B: Full set of conditions: Stimulus-response conditions & mental decision-making action

$\# Combinations = 2^{\sum C_1 C_2 \ldots C_n} = 2^3 = 8$

Title Does the employee satisfy the conditions, discover the link beneath the instructions paragraph, and continue?

#	Conditions (C)	# Combinations 1	2	3	4	5	6	7	8
C1	((employee.date of birth >= 1971) OR (employee.joined retirements plan > 20100101)) AND (submission =<20111215)	Y	Y	Y	Y	N	N	N	N
C2	the employee sees the link "→ Choosing the contribution option in the savings plan", assumes that the link will lead to task accomplishment	Y	Y	N	N	Y	Y	N	N
C3	the employee decides to set contribution options (mental action)	Y	N	Y	N	Y	N	Y	N
	Deductive scenarios:								
	Reasonably possible scenarios	X	X	X	X		X		X
	Error scenario			X1		X		X	
	Actions								
A1	read instructions; discover link "→ Choosing the contribution option in the savings plan"; clicks link "→ Choosing the contribution option in the savings plan";	X							
A2	cancel (employee does not want to change contribution options); reads OR does not read the rest of the page;		X		X		X		X
A3	read instructions; the employee decides to search for "MyHR+"; the employee needs to store the instructions given in working memory but the instructions are too complicated to keep in working memory; the search for "MyHR+" can actually be done on the page where the employee is OR the employee goes back to the default "home" page (scenario not modelled; if done, search time will increase);			X		X		X	
X1	**Leap of faith:** design made without evidence of human behavior								

= reasonably possible scenario by combination

Figure 3 Decision tables. A: 2 Pre-conditions (left). B: 2 Pre-conditions and 1 mental action (right)

The design of our case study has not considered any emotional-motivational aspects of human activity like probability of task strategy, significance, changing goals of activity, DP, and more. However, our findings from QA enable us to propose an ameliorated method of task performance, which overcomes shortcomings from our discoveries during QA. However, the extent of this paper constrains us from an in-depth discussion, our approach based on human goal-oriented activity towards the solution space in Figure 4 greatly simplifies task complexity, reduces cognitive overload, resolves negative emotional-motivational aspects, responds to task significance, and is now economical. After all, without our painstaking semi- and formal QA within the holism of human goal-oriented activity, the leap of faith would still be shrouded.

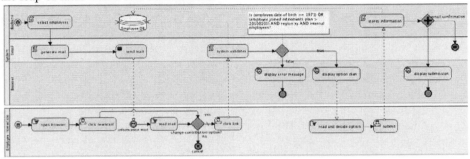

Figure 4 Proposed new method of task performance (pooled and layered; excludes probabilities)

5 CONCLUSION & FUTURE RESEARCH

We have presented a practical scenario of an apparently trivial and everyday browsing behavior on the corporate intranet of a human subject (HS) whose initial goal of activity with low motivational task significance has instantaneously changed to a new goal of activity with high significance. Contrary to his subjective goal and the motivational significance that task strategy reveal, the objective goals of the design entirely differ and resemble a leap of faith: Although our motivated HS succeeded in submitting his next year's contribution of the local pension fund after a long navigational maze with an insidiously growing frustration, his effort has been in vain because he never qualified for the action yet was never informed by the system. Individual significance hence closely affects the interaction between cognitive and emotional-evaluative mechanisms that mutually stimulate a HS to select and interpret information presented at the computer screen. This phenomenon influences the cognitive strategy of activity when HS interact with computers and might explain why HS accept and still pursue inadequate goals of activity for a specific task.

By the same token, the holism of human goal-oriented activity invites design to no longer ignore the vector 'motive(s)→ goal' of an individual HS. This vector is the point of departure for subjective significance, may elevate a new goal of activity, affects decision-making in goal-acceptance, execution, and the evaluation of a result, and determines probabilities of task strategies that proceed from decision points (DP). Our practical application using qualitative analysis (QA) of the Systemic-Structural Activity Theory proves interchangeable relationships of emotional-motivational aspects within the hierarchical order of human activity and behavior in time and space. As such, our findings extend those boundaries of the mere stimulus-response behavior and the narrow views of design in Software Engineering and Human-Computer Interaction by our rationalization to incorporate significant and identifiable mental conditions from the universe of all possible task strategies into DP of the subject domain for any work process or man-machine system i.e., everyday' and safety-critical systems alike. Our approach and the understanding of DP are critical in e.g., error analysis, accuracy, simplification, prediction, simulation, and replications of human behavior and task strategies during goal-oriented human work activity. Contrary to the original design of our actual case, our QA of *one* individual HS suggests a solution that simplifies task complexity, reduces cognitive overload, resolves negative emotional-motivational stimuli, responds to task significance, and is economical.

Future research will algorithmically analyze the erroneous behavior in a graphic-symbolic description and classification of actions. Coupled with quantitative aspects, we will examine probabilities, DP and their types, determine the classes of algorithm that cause subjective and objective misbehaviors, abstract and compute abandoned actions, assess task reliability of the existing and new methods of task performance, and draw conclusions about the efficiency of the proposed improvement.

REFERENCES

Abelson, R.P., K.P. Frey and A.P. Gregg (Eds.). 2004. *Experiments With People: Revelations From Social Psychology* Lawrence Erlbaum, Mahwah, NJ.

Anokhin, P.K. 1966. Special Features of the Afferent Apparatus of the Conditioned Reflex and Their Importance to Psychology. In *Psychological research in the U.S.S.R.* Eds. Leontyev, A., A. Lurya and A. Smirnov. Progress Publishers, Moscow: pp. 67-98.

Bedny, G. and W. Karwowski 2007. *A Systemic-Structural Theory of Activity: Applications to Human Performance and Work Design*. Taylor & Francis, Boca Raton.

Bedny, G., H. von Brevern and K. Synytsya. 2012. Learning & Training: Activity Approach. In *Encyclopedia of the Sciences of Learning*. Ed. Seel, N.M. Springer, Boston, MA: pp. 1800-1805.

Bedny, I.S. and G.Z. Bedny. 2010. Abandoned Actions Reveal Design Flaws: An Illustration by a Web-Survey Task, #6. In *Human-Computer Interaction and Operators Performance - Optimizing Work Design with Activity Theory*. Eds. Bedny, G. and W. Karwowski. Taylor & Francis, pp. 149-182.

Synytsya, K. and H. von Brevern. 2010. Information Processing and Holistic Learning and Training in an Organization: A Systemic-Structural Activity Theoretical Approach, #15. In *Human-Computer Interaction and Operators Performance - Optimizing Work Design with Activity Theory*. Eds. Bedny, G. and W. Karwowski. Taylor & Francis, pp. 385-409.

von Brevern, H. and K. Synytsya. In press. Towards 2020 with SSAT. In *Science, Technology, Higher Education and Society in the Conceptual Age*. Eds. Marek, T., W. Karwowski and J. Kantola. Taylor & Francis, London, New York: pp. tbd.

CHAPTER 15

Using Systemic Approach to Identify Performance Enhancing Strategies of Rock Drilling Activity in Deep Mines

Mohammed-Aminu Sanda[1,2], Jan Johansson[1],
Bo Johansson[1], Lena Abrahamsson[1]

[1]Luleå University of Technology
Luleå, Sweden

[2]University of Ghana Business School
Legon, Accra, Ghana
masanda@ug.edu.gh, jan.johansson@ltu.se,
bo.johansson@ltu.se, lena.abrahamsson@ltu.se

ABSTRACT

This paper looked at the need for understanding the sociotechnical and psychosocial characteristics of rock drilling activity in deep mines that could lead to the harmonization of the human, technological, and organizational components of the work systems. The aim is to identify performance enhancing strategies that could be used to improve and optimize human-technology collaboration in rock drilling activity in deep mines. Guided by the systemic structural activity theory, data was collected by video recording two skilled miners engaged in two separate rock drilling activities using a high technology drilling machine at a Swedish underground mine. Using the systemic analytical approach, the data obtained were analyzed morphologically and functionally. Results from analysis of the miners' motors actions during the rock drilling activity showed that by using implicit driven strategies, they were able to perform simultaneously two specific tasks that required high levels of concentration and visual control in the normal visual field available to them from inside the protective cabin of the high technology equipment they were using. They simultaneously combine their mental actions and motor actions in recognizing and remedying the constraining effects of unfamiliar stimuli during the rock drilling activity. It is concluded that the functional efficiency and effectiveness of rock drilling activity as well as the miner's productive performance in future automated and digitized deep mines could be enhanced by identifying the implicit characteristics of their performance enhancing actions and operational strategies. This understanding has future implications in designing a very efficient and effective human-technology collaboration in a highly digitized deep mine work system.

Keywords: rock drilling activity, performance enhancing strategy, work system, systemic analysis, digitized deep mine

1   INTRODUCTION

There is an ongoing discussion about the design of production systems for future mines based on technological development for measurement and process management, organizational design and learning for employees. Since such systemic processes are managed and monitored by people, it is important to harmonize the functional roles of the technical and social components of the system. This therefore calls for the need to develop knowledge to facilitate the integration of technological, organizational and human systems. The sense here is that the developmental approach for the intelligent automation system should not be viewed only from the perspective of designing systems/automation adaptable to humans. Humans should rather be considered as integral resources whose integration can enhance the possibility of designing better systems (intelligent automation). In this respect, therefore, it is important to develop knowledge that can enhance the organization of work in future mines. In other words, there is the need to find a possible way forward for the mining industry rationalization by focusing on work organization where cooperation, skills development and learning are key components. Such need is derived from the challenges posed by the prevailing thinking among firms in the mine industry that by replacing humans with machines at all levels in the value chain, a rapid increase of automation and integration of various processes and unit operations can occur to enhance the economic viabilities and competitiveness of mining firms in the future. This observation has also brought to the fore the realization that increased automation, combined with global competition, is leading mining firms to rely on a lean organization with multi-skilled workers capable of managing multiple areas of the production activity which requires people and machines to cooperate with each other. The rationale is to make the mining work environment a substantially different place in the future.

Possible work environment goals for automated production systems must have the capacity to support a high situation awareness that stimulates good performance, and at the same time establish a reasonable perceived subjective workload. Such production system should neither over-stimulate nor under-stimulate operators to over-rely on decisions made by computers. It should also not create automation bias in terms of an over-reliance on decision support from the computer system, but must provide reliable operator feedback to prevent possible mistakes. Thus there is the need for the development of knowledge to facilitate the integration of technological, organizational and human systems. The sense here is that the developmental approach for the intelligent deep mining should not be viewed only from the perspective of designing systems adaptable to humans. Humans should rather be considered as integral resources whose integration can enhance the possibility of designing better systems.

2 LITERATURE REVIEW

Technologies are used in organizations for the accomplishment of work activities. In this respect, technologies can be deemed to affect social relations in organizations. Aldrich (2008) related such technological effect on social relations to the technology's capacity to structure transactions between roles in an organization's activity system. In this respect, it is important for the mining industry to know that extensive automation create inflexible, rigid, expensive and complex solutions. Arguing along the lines of Bedny and Karwowski (2007), this necessitate the need to create a balance between technical resources and human work in situations where the production component of the organizational activity system is characterized by many and rapid changes. The complexity of the human work process is highlighted by the characteristics of its substructure whose basic components have been delineated by Bedny and Karwowski (2007) as follows: (i) motive-goal as a vector which demonstrates the directional and energetic aspects of the work activity, (ii) knowledge and skills which demonstrate the relevance past experience to the work process, (iii) abilities related to the tasks to be performed, and (iv) work actions which are organized into a structure, and together present the method of work. Action in this sense implies both cognitive and motor action.

The presence of the concepts of knowledge and action in the structure of work process also implies the existence of mental tools (Bedny and Karwowski, 2007). Therefore, in the process of developing new work practices in an organization, it is important to see the key actors in the practice development exercise as learners of their new activities (Sanda, Johansson and Johansson, 2011). The implication here is that when employees are not actively involved throughout the planning and implementation processes of the work practice, it often results in a poorly designed work system and a lack of employee commitment (Sanda, 2006). This is because in the task interpretation process, the worker has to be able to involve his personal prerequisites such as experience, skills and physical constitution, as well as his/her context as part of social systems inside and outside the organization (Sanda, 2008). Additionally, the worker has to solve all the problems that are not taken care of or misinterpreted when management design tasks (Hendrick and Kleiner, 2001). This observation is indicative of Lave and Wenger's (1991) argument that in a 'community of practice' individual thought is essentially social and is developed in interaction with the practical activities of a community, through living and participating in its experiences over time. Thus in future deep mining activity, Bardram's (1997) characterization of an activity as consisting of several actions, and each action also consisting of several operations attains a sense of significance (Sanda et al., 2011). This is more so when the notion of activity is considered from the systemic structural activity theoretical perspective (Bedny and Karwowski, 2007) as an organized system with a discrete structure. By this, the organizational activity of deep mining could be categorized into three levels. These are the activity level, the action level and the operation level. Based on this categorization, an organizational activity is a system consisting of chains of actions, which in turn consist of chains of operations/tasks. Generally, the more complex a task orientation, the more the mental efforts required for its performance (Bedny and Karwowski, 2007). Bedny and Karwowski (2007) recommend the following major criteria for the classification of tasks; (i) indeterminacy of initial data, (ii) indeterminacy of tasks, (iii) existence of redundant and unnecessary data for task performance (iv) contradictions in task conditions, and complexity or difficulty of task (vi) time restrictions in task performance (vii) specifics of instructions, and their ability to describe adequate performance and restrictions, and (viii) adequacy of subject's past experience for task requirements.

3 METHOD AND MATERIAL

3.1 Data Collection

Data was collected in a Swedish underground mine. The data collection entailed the conduction of both morphological and functional studies of two highly-skilled miners engaged in the rock drilling activity. In this regard, the external behavioral and internal mental actions and operations of the miners in the rock drilling activity were observed and video-recorded. The external behavioral actions include various motions that are used to transform material or tangible objects. The mental actions transform images, concepts or propositions and non-verbal signs in the mind (Bedny and Karwowski, 2011). In the data collection procedure, miners' engaged in two separate rock drilling activities using the Boomer, which is a highly automated machine with two robotic drilling arms, were video recorded.

3.2 Data Analysis

Using the systemic analytical approach as the basic paradigm for the analysis of positioning actions (Bedny and Karwowski, 2011) in rock drilling activity in deep mines, both morphological and functional analyses were conducted. In the morphological analysis, the constructive features of the rock drilling activity, entailing the logical and spatio-temporal organization of the cognitive behavioral actions and operations involved, were described. In the functional analysis, potential strategies of activity performance associated with the miners' actions and their corresponding operations, identified as constituting as functional blocks were analyzed qualitatively using systemic principles (Bedny and Karwowski, 2007). This allowed for the evaluation of varieties of performance indicators, such as time and errors, and also the selection of the most efficient strategy.

4 RESULTS

Observations and video recordings of the miners' engagement with the production drilling activity show that a miner's individual object-oriented activity consists of both physical and mental actions and operations whose characteristics are influenced by past experiences. The functional appraisal of this activity shows that the object of a miner's self-regulation of orienting activity to be two folds. The object, in terms of the miner's physical activity, is the conduction of production (rock) drilling operations, based on informed decisions and programs to orient explorative actions towards performance enhancement, sensemaking in task performance, and determination of motivational level. In terms of the miner's mental activity, the object is the simultaneous observation of the production drilling work, assessment of task difficulty, listening to communication models in order to enhance the development of stable or dynamic mental models of task or situation for enhancing the relevant task conditions. The object also includes the setting of subjective standard of "successful results" and also "admissible deviation".

Morphological analysis of the miners' mental assessment of task difficulty showed that; they view the use of tractor technology with more than one robotic arm (i.e. boomer) for production drilling tasks as excessive for one operator to handle. The miners' related the perceived task overload to the difficulty an operator encounters in his/her ability to focus on the computerized programming command for the automated rock drilling actions and operations using the multi boomers. The miners viewed such situation as distracting them from developing the requisite dynamic mental models of task or situation which are required for enhancing the quality of the relevant task conditions. The miners view that even in situations where high technology robotic loaders that are remotely controlled from safe distances are used for important task components in production drilling activity, the guide cameras used to manipulate the robots' movements do not guide their operations as efficiently as the direct use of the operator's human eyes. As such the operators, based on acquired experiences, mostly find ways to guide the technology for optimum performance. A sense of this was provided qualitatively by the miners as follows:

> The technology does not always get it right. Most of the time, I use the experience I have acquired over the years to guide the technology for optimum performance. I have also developed enormous knowledge on manipulating the technology to make my work activity easier.

Functional analysis of the scenario above shows that the miners possess tacit knowledge developed overtime on various objective activities, but which remained shared. This tacit knowledge is used by the operators to negotiate technology-based standardized task patterns in bids to overcome task repetitiveness and also to increase their productive capacities in terms of waste removal in production time. The miners also use their tacit knowledge to overcome subjective perceptions of technological shortcomings in their task undertakings (based the notion that technologies do not always get it right). In some instances, in the bolting operations, operators negotiate tasks to make them move faster. For example, it was found that operators of the high technology machine in the roof drilling and bolting operations in the deep mine used techniques enhanced by old mining culture to negotiate and accelerate the tasks. Instead of the operator following the sequential roof bolting pattern of firstly drilling a hole, followed by pumping cement in the drilled hole, and then inserting bolted rod in the cement-filled hole to complete task (i.e. hole drill + cement + insert bolted rod), the operator negotiate this pattern by firstly drilling as many holes as possible before pumping cement and then inserting bolted rod in each of the drilled holes. An illustration of this scenario entailing the sequential roof bolting pattern and the operator's negotiated pattern for three roof bolting tasks and their respective durations for completion are presented in table 1 and table 2 below.

Table 1: Task analysis of Operator's standardized pattern for completing three roof bolting activities.

Task	Sub-Tasks	Task Duration
1	Drill hole 1 in rock ↓ Fill cement (grout) in hole 1 ↓ Insert iron rod in cement-filled hole 1 to complete task	5 min 38 sec
2	Drill hole 2 in rock ↓ Fill cement (grout) in hole 2 ↓ Insert iron rod in cement-filled hole 2 to complete task	5 min 36 sec
3	Drill hole 3 in rock ↓ Fill cement (grout) in hole 3 ↓ Insert iron rod in cement-filled hole 3 to complete task	5 min 40 sec
	Time for Activity Completion	16 min 54 sec

Table 2: Task analysis of Operator's negotiated pattern for three roof bolting actions.

Task	Sub-Tasks	Task Duration
1	Drill hole 1 in rock. ↓ Drill hole 2 in rock. ↓ Drill hole 3 in rock.	7 min 10 sec
2	Fill cement (grout) in hole 1. ↓ Insert iron rod in cement-filled hole 1 to complete task.	2 min 16 sec
3	Fill cement (grout) in hole 2 ↓ Insert iron rod in cement-filled hole 2 to complete task	2 min 15 sec
4	Fill cement (grout) in hole 3 ↓ Insert iron rod in cement-filled hole 3 to complete task	2 min 16 sec
	Time for Activity Completion	13 min 57 sec

As it is shown in table 1, the standardized procedure requires the completion of the hierarchical operational level for each roof bolting task, before the next task commences. Overall, three task steps are required. Hence the duration for performing all the three roof bolting activity is the summation of the total operational time for each of the three tasks, which is 16 minutes, 54 seconds. For the negotiated pattern shown in table 2, the first level operational task of rock drilling is repeated consecutively for all the three sub-tasks. This is then followed by the completion of sub-tasks for levels 2 and 3 for each drilled hole in each of the three roof bolting activity. Overall, four tasks steps are required. Hence the duration for performing all the three roof bolting tasks is the summation of the total operational time for each of the four tasks, which is 13 minutes, 57 seconds.

Comparison of the duration outcomes in table 1 and table 2 shows a time reduction of 2 minutes 57 seconds in the miners' negotiated task performance of the three bolting tasks (i.e. from 16 minutes 54 seconds to 13 minutes, 57 seconds). This finding confirms the miner's notion of time reduction in performing same task as a result of operators' use of tacit knowledge to negotiate the task pattern in the roof bolting activity.

5 DISCUSSION

Based on the analysis in section four above, we argue that since organizations possess technologies (i.e. techniques for processing raw materials and/or people) for accomplishing work, organizational activity then emphasizes a work system design in which technology affects social relations in organizations by structuring transactions between roles that are building blocks of an organization. In this respect, we argue that application of systemic-structural activity theory stands to provide an understanding of the various processes that is entailed in digitized human work which can be used to design a harmonious work environment integrating the human, technical and the social system, towards increased productivity in the deep mine industry. The findings from the morphological analysis have shown that relationship of between the miners' external behavior and internal psychological functions are mutually regulated. Arguing from the perspectives of Bedny and Karwowski (2007), we find this relationship to entail a process of internalization in a miner's motor activity, which is central to the theory of activity. By implication, we find a degree of commonality in the cognitive characteristics of the miners' internal mental activity and the regulative nature of their external behaviors towards activity undertakings. This finding is underscored by the argument supporting the mutual interdependence of mental development, semiotic mediation and external practical activity which do not exist separately (Bedny and Harris, 2005; Bedny and Karwowski, 2007). The sense here is that an inter-subjective aspect of activity is observable in a miner's individual activity.

Findings from the functional analysis support the miners' notion of time reduction in same task undertakings using tacit knowledge to negotiate an action pattern in an activity. It also portrayed the structure of the miners' activity during task performance as a logically organized system of cognitive and motor actions and operations (Bedny, Karwowski and Sengupta, 2008) that enhances innovation. This shows that the specificity of the rock drilling activity is underlined by the interdependence of the miner's practical activities and symbolical activities, each of which is in constant transformation of the other. Since the practical actions in the rock drilling activity have clearly defined object, it then entails semiotic mediation. The occurrence of such mediation is highlighted by the conscious goal (including planning and understanding of the possible outcomes) with which the object-practical actions in the rock drilling activity are undertaken.

Despite the miners' use of highly reliable mobile mining equipment with improved operating and maintenance procedures, the relevance of their deep knowledge about the rock (i.e. their skills and ability to 'read the rock') molded by the old mining culture comes into play as a result of process interruption when a hard rock is encountered. The development of the tacit knowledge was to cater for the miners' observations that the technologies they use in their task undertakings do not always get it right. All these presuppose the presence of symbolic representation of reality (Bedny and Karwowski, 2007). In this regard, we argue that internal activity rock drilling is constructed by the miners based on the mechanisms of self-regulation. Thus, for the miners engaged in roof bolting activity, their active exposure diversity derived from years of practice and experience appeared to have increased the number of conceptual categories (Hendrick, 1991) they have developed for storing information. They also appeared to have tacitly developed new rules and combinations of rules for integrating conceptual data in the actions and operations associated with roof bolting, and as a result provided more insightful knowledge of the complex problems and solutions in the roof bolting activity. The implication here is that the miners, being cognitively complex persons tend to be open in their beliefs and relativistic in their thinking, as well as have a dynamic conception of their work environment. As Abrahamsson and Johansson (2008) argues, skills and knowledge should not be seen as things that are simply static and accumulated by individuals, but rather as things that are created and changed in socio-cultural contexts, through individual as well as collective processes. Thus from the practical point of view, we ascribe to Bedny and Karwowski's (2007) position that the interrelationship of internal and external activity in rock drilling determines what the miner's practical-external activity are, and also the external tools for the successful course of the miner's mental activity. We view the study of this interrelationship to be important for the future formulation of training methods in rock drilling activity. This is because newer technologies are now bringing other business processes into the frame of mining activity with smaller workforce having expanded role responsibilities

6 CONCLUSION

The results and the corresponding discussion in sections four and five above have shown that the object-practical activity of the rock drilling activity in the deep mines is determined by the genesis and content of the miners' mind. In this respect, therefore, we conclude that the functional efficiency and effectiveness of the production drilling activity in future automated and digitized deep mines could be increased by identifying performance enhancing strategies that are used by workers to facilitate the social collaboration between them and the technology they use in order to their productivities. Identification of such performance must require an understanding of the interrelationship of internal and external activity which stands to determine the miners' practical-external activity and the corresponding external tools that they need to enhance their mental activities towards developing successful performance enhancing strategies in the rock drilling activity. By implication, understanding derived from such performance enhancing strategies could be integrated in the design of efficient and effective work systems and/or technology.

REFERENCES

Abrahamsson, L. and Johansson, J. 2008. Future mining: Workers' skills, identity and gender when meeting changing technology. In: *Future mining,* ed. S. Saydam. Proceedings of the First International Future Mining Conference and Exhibition. Carlton Victoria, New South Wales, Australia: The Australasian Institute of Mining and Metallurgy: 213-220.

Aldrich, H. E. 2008. *Organizations and environments*. Stanford, California: Stanford University Press.

Bardram, J. E. 1997. Plans as situated action: An activity theory approach to workflow systems. *Proceedings of the Fifth European Conference on Computer Supported Cooperative Work*: 17-32.

Bedny, G. Z. and Harris, S. R. 2005. The systemic-structural theory of activity: applications to the study of human work. Mind, Culture, And Activity, 12 (2): 128-147.

Bedny, G. Z. and Karwowski, W., 2007. *A Systemic-structural theory of activity: Applications to human performance and work design*. Boca Raton: CRC Press.

Bedny, G. Z. and Karwowski, W. 2011. Analysis of strategies employed during upper extremity positioning actions. *Theoretical Issues in Ergonomics Science*, iFirst: 1–20.

Bedny, G. Z., Karwowski, W. and Sengupta, T. 2008. Application of Systemic-Structural Theory of Activity in the Development of Predictive Models of User Performance. *International Journal of Human-Computer Interaction*, 24 (3): 239-274.

Hendrick, W. 1991. Human factors in organizational design and management. *Ergonomics*, 34: 743-756.

Hendrick, H. W. and Kleiner, B. M. 2001. *Macroergonomics: An introduction to work system design*. Santa Monica: HFES.

Lave, J. and Wenger, E. 1991. Situated *learning: Legitimate peripheral participation*. Cambridge, UK: Cambridge University Press.

Sanda, M. A. 2006. *Four case studies on the commercialization of Government R&D agencies: An organizational activity theoretical approach*. Luleå, Sweden: Luleå University of Technology Press.

Sanda, M. A. "Exploring the concept of emerging object of activities in understanding the complexity of managing organizational change". Paper presented at the Second ISCAR Conference. San Diego, California, 2008.

Sanda, M. A., Johansson, J. and Johansson, B. 2011. Conceptualization of the attractive work environment and organizational activity for humans in future deep mines. *International Journal of Human and Social Sciences*, 6 (2): 82-88.

Section IV

Cognitive Engineering: Error and Risk

CHAPTER 16

Challenges in Human Reliability Analysis (HRA): A Reflection on the Accident Sequence Evaluation Program (ASEP) HRA Procedure[1]

Huafei Liao[1], Alysia Bone[2], Kevin Coyne[2], John Forester[1]

[1] Sandia National Laboratories, USA
[2] U.S. Nuclear Regulatory Commission, USA

ABSTRACT

Human reliability analysis (HRA) is used in the context of probabilistic risk assessment (PRA) to provide risk information regarding human performance to support risk-informed decision-making with respect to high-reliability industries. In the current state of the art of HRA, variability in HRA results is still a significant issue, which in turn contributes to uncertainty in PRA results. The existence and use of different HRA methods that rely on different assumptions, human performance frameworks, quantification algorithms, and data, as well as inconsistent implementation from analysts, appear to be the most common sources

---

[1] The opinions expressed in this paper are those of the authors and not those of the U.S. Nuclear Regulatory Commission (NRC) or of the authors' organizations. This report was prepared as an account of work sponsored by an agency of the U.S. Government. Neither the U.S. Government nor any agency thereof, nor any of their employees, makes any warranty, expressed or implied, or assumes any legal liability or responsibility for any third party's use, or the results of such use, of any information, apparatus, product, or process disclosed in this report, or represents that its use by such third party would not infringe privately owned rights.

for the issue, and such issue has raised concerns over the robustness of HRA methods. In two large scale empirical studies (Bye et al., 2012; Forester et al., 2012), the Accident Sequence Evaluation Program (ASEP) HRA Procedure, along with other HRA methods, was used to obtain HRA predictions for the human failure events (HFEs) in accident scenarios. The predictions were then compared with empirical crew performance data from nuclear power plant (NPP) simulators by independent assessors to examine the reasonableness of the predictions. This paper first provides a brief overview of the study methodology and results, and then discusses the study findings with respect to ASEP and their implications in the context of challenges to HRA in general.

Keywords: human reliability analysis (HRA), ASEP, THERP, probabilistic risk assessment (PRA), crew performance

1 INTRODUCTION

Human reliability analysis (HRA) can be defined as the use of systems engineering and behavioral science methods in order to render a complete (as possible) description of the human contribution to risk and to identify ways to reduce that risk. It is used in the context of probabilistic risk assessment (PRA) to provide risk information regarding human performance to support risk-informed decision-making with respect to high-reliability industries. For example, risk information from HRAs is an important input to the U.S. Nuclear Regulatory Commission (NRC) for their licensing and regulatory decisions.

In the current state of the art of HRA, variability in HRA results is still a significant issue, which in turn contributes to uncertainty in PRA results. The use of different HRA methods that rely on different assumptions, human performance frameworks, quantification algorithms, and data, as well as inconsistent implementation from analysts, appear to be the most common sources for the issue, and such issue has raised concerns over the robustness of HRA methods.

The Accident Sequence Evaluation Program (ASEP) HRA Procedure (referred to as "ASEP" in this article) (Swain, 1987) is based heavily on the Technique for Human Error Rate Prediction (THERP) (Swain and Guttman, 1983) method with simplified human performance models and guidance, and it was developed to enable systems analysts at reasonable cost, with minimum support and guidance from experts in HRA, to make estimates of human error probabilities (HEPs) and other human performance characteristics that are sufficiently accurate for many PRAs. The analysts essentially quantify HEPs by first estimating the response times required to perform some critical actions and evaluating factors prescribed by the ASEP guidance and relevant to the human failure events (HFEs) being addressed. Then, they select appropriate HEPs (with uncertainty bounds) from the tables and curves provided in ASEP based on the assessment of the factors. Like THERP, ASEP relies on a time reliability curve (TRC) for quantifying the probability of failure in the diagnosis portion of human actions. Although ASEP allows use of

THERP to support quantification of post-diagnosis actions, ASEP is almost entirely self-contained as an HRA method; the analysts do not need to be familiar with THERP nor are they required to use any of the THERP models or data.

2 OVERVIEW OF THE STUDY METHODOLOGY AND RESULTS

Two large scale empirical studies were conducted to develop an empirically based understanding of the performance, strengths, and weaknesses of different HRA methods used to model human response to accident sequences in PRAs. The empirical basis was developed through simulator runs with real crews responding to accident situations similar to those modeled in PRAs. The two studies are referred to as the International HRA Empirical Study (Lois et al., 2009; Bye et al.,2011; Dang et al, 2011; Forester et al., 2012) and the U.S. HRA Empirical Study (Bye et al., 2012; Marble et al., 2012), and were multinational collaborative efforts involving organizations from 10 countries. The U.S. NRC, in particular, played a major role in supporting the preparation and execution of the studies.

In both studies, two teams employed ASEP to obtain HRA predictions for the HFEs in the accident scenarios. The predictions were then compared with the crew data by independent assessors to examine the accuracy of the predictions. The aggregated crew performance is described in the following three ways, which correspond to those in which the HRA teams were asked to report their predictions and serve as the data for comparing with the HRA predictions.

- Performance on the HFE related actions expressed in operational terms ("operational descriptions");
- Assessment of the PSFs (main drivers) for each action;
- Number of crews failing to meet the success criteria for each action and an assessment of the difficulty of the action.

The comparisons examined both the qualitative and quantitative method predictions. Qualitative predictions include, for instance, the aspects of the scenario or task conditions identified as the driving factors influencing operating crew performance in responding to the scenario. The quantitative comparisons take into account the estimated failure probabilities of the defined HFEs of interest and their correspondence with the observed difficulty of the HFEs (Bye et al., 2012; Forester et al., 2012).

In the International Study, two categories of scenarios were performed: steam generator tube rupture (SGTR) and loss of feedwater (LOFW) scenarios. Nine HFEs were defined in the SGTR scenarios and four HFEs in the LOFW scenarios. Fourteen European nuclear power plant (NPP) operator crews participated in the study, and their performance data were collected in the Halden Reactor Project's HAMMLAB (HAlden huMan-Machine LABoratory) NPP simulator facility (Lois et al., 2009; Bye et al.,2011; Dang et al, 2011). It should be noted that some crews did not complete the LOFW scenarios due to simulator problems.

In the US Study, there were three scenarios. Scenario 1 was a total LOFW followed by an SGTR, for which three HFEs were defined. Scenario 2 was a loss of component cooling water (CCW) and reactor cooling pump (RCP) sealwater, for which one HFE was defined. Scenario 3 was an SGTR scenario without further complications, in which one HFE was defined. Four crews from a participating U.S. NPP participated in the U.S. Study, and their performance data of four HFE scenarios were collected on the plant full-scope training simulator (Bye et al., 2012; Marble et al., 2012). It should be noted that one crew was unable to complete Scenario 3 (SGTR) due to a simulator problem.

The HFEs were ranked in terms of problems experienced by the crews in diagnosing and completing the actions based on the empirical data, as well as by opinions of the crew members who participated in the study. The crew failure rates and difficulty rankings for both studies are listed in Tables 1 to 3. It should be noted that HFEs 5B1 and 5B2 in the SGTR scenarios are exclusive, meaning that data were available for only one HFE for a particular crew.

Table 1. Crew Failure Rates and HFE Difficulty Ranking for the SGTR Scenarios in the International HRA Empirical Study

HFE	Failure Rate	Difficulty Ranking
HFE 5B1	7/7	5 (Very difficult)
HFE 1B	7/14	4 (Difficult)
HFE 3B	2/14	3.5 (Somewhat difficult)
HFE 3A	1/14	3 (Somewhat difficult)
HFE 1A	1/14	2.5 (Easy to Somewhat difficult)
HFE 2A	1/14	2.5 (Easy to Somewhat difficult)
HFE 2B	0/14	2.5 (Easy to Somewhat difficult)
HFE 5B2	0/7	2 (Easy)
HFE 4A	0/14	1 (Very easy)

Table 2. Crew Failure Rates and HFE Difficulty Ranking for the LOFW Scenarios in the International HRA Empirical Study*

HFE	Failure Rate	Difficulty Ranking
HFE 1B	7/10	5 (Very difficult)
HFE 2B	0/7	3.5 (Somewhat difficult to difficult)
HFE 1A	0/10	2.5 (Easy to somewhat difficult)

*Empirical data were not available for HFE 2A.

Table 3. Crew Failure Rates and HFE Difficulty Ranking for the Scenarios in the US HRA Empirical Study*

HFE	Failure Rate	Difficulty Ranking
HFE 2A	4/4	1 (Very difficult)
HFE 1C	3/4	2 (Difficult)
HFE 1A	0/4	3 (Fairly difficult to difficult)
HFE 3A	0/3	4 (Easy)

*Empirical data were not available for HFE 1B.

Alongside the Bayesian uncertainty bounds derived from the crew data, the HEPs predicted by the two analyst teams are plotted in Figures 1 and 2 for the SGTR and LOFW scenarios, respectively. On the horizontal-axes, the HFEs are ordered by their difficulty ranking.

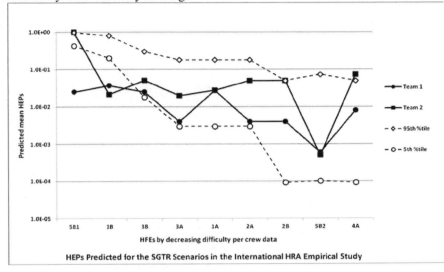

Figure 1. Predicted Mean HEPs with Bayesian Uncertainty Bounds for the SGTR Scenarios in the International HRA Empirical Study

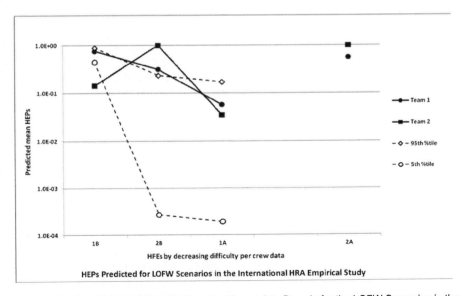

Figure 2. Predicted Mean HEPs with Bayesian Uncertainty Bounds for the LOFW Scenarios in the International HRA Empirical Study

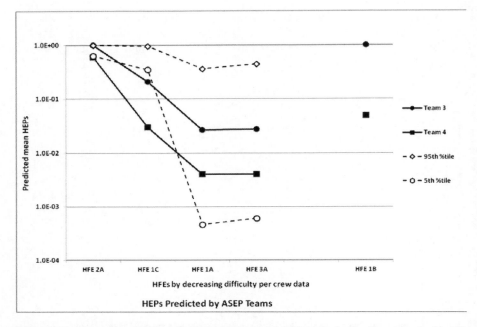

Figure 3. Predicted Mean HEPs with Bayesian Uncertainty Bounds for the Scenarios in the US HRA Empirical Study

3 INSIGHTS FROM THE STUDIES

In the two studies, ASEP was assessed to explore where the method itself appeared to be contributing to strengths or shortcomings in the predictive ability of the analysis, particularly in terms of the guidance provided in the method, and where analysts were varying in their implementation of the method (e.g., by going beyond the method guidance or not using the method as designed). The strengths and weaknesses for ASEP are summarized below.

3.1 Strengths

1) Simplicity

One strength of ASEP is ease of use due to its simplicity, such as simplifying its human performance model by separating diagnosis from post-diagnosis actions, estimation of the diagnosis HEP only with the TRC with a few PSF adjustments, and focus only on the major procedural steps without examining potential complexities in the sub-steps given the conditions of the scenario. On one hand, the simplifications make the method easy to use; on the other hand, they seem to contribute to the weaknesses discussed below. The simple analysis is justified by the developer by claiming that conservative HEPs will generally be obtained. However, apparent optimism due to the method's weaknesses was seen in some cases (e.g., the analysis of HFEs 1B and 5B1 in the SGTR scenarios by Team 1 in the International Study) (see discussion below). The implication is that the tradeoffs between simplicity and thorough analysis need to be weighed before applications of the method.

2) Traceability

At some level, another strength of the method is its traceability. The estimation of allowable diagnosis time and allowable post-diagnosis time, the derivation of the HEP within the method, and what is important to performance given the factors considered is generally traceable, and how the various factors are weighted in determining the final HEP can be determined. However, how analysts might bias or alter the rating or level of the factors considered in applying the method, based on other information identified that is not covered by the method would be difficult to trace if the analysts do not document their decision process well.

3.2 Weaknesses

1) Insufficient guidance on when to consider cognitive demands in connection with the execution of a task.

By segmenting total time available for coping with an abnormal event into two artificially independent parts: *allowable diagnosis time* and *allowable post-diagnosis time (i.e., related to response execution time)*, ASEP provides an option to explicitly include and quantify diagnosis or not. However, insufficient guidance is provided as to when to include or exclude diagnosis. In the International Study, Team 1 assumed that no diagnosis was required once the crews entered symptom-based procedures in SGTR scenarios, and Team 2 made such an assumption in both SGTR and LOFW scenarios. The crew data showed that such an assumption was not well justified as crews had to assess the situations and/or make new response plans while the scenarios progressed. Failure to address diagnosis seemed to be a major contributing factor to the team making predictions inconsistent with the empirical data in terms of performance drivers and operational stories. Apparently, the decision to skip the diagnosis part of crew response may have precluded the opportunity to address operators' cognitive activities, examine the difficult conditions operators would be facing, and identify some important factors influencing performance. As a consequence, the HRA teams only obtained a partial picture of the dynamic nature of the accident scenarios, which was seen in both teams' analyses in the International Study. For example, by focusing mainly on crews working through the procedures, Team 1 in the International Study did not really distinguish the strong difference between the conditions for HFEs 5B1 and 5B2 in the SGTR scenarios. In addition, except for the easiest HFEs 5B2 and 4A where there seems to be a good agreement between the predicted drivers and those identified from the crew data, the predicted negative drivers rarely matched those identified from the crew data in the SGTR scenarios. In contrast, the team identified many of the important drivers that would influence performance in the LOFW scenarios. Although this seems to be partly due to team's experience from conducting HRAs for the SGTR scenarios (there might be a learning effect for Team 1 in the LOFW scenarios, because HRAs for the LOFW scenarios were conducted after the team saw the study results for the SGTR scenarios) an effort that went beyond ASEP guidance, addressing diagnosis in terms of the ASEP TRC did lead the team to obtain a good understanding of what would be going on in the scenarios and consider the potential impact of time available on the diagnosis.

Quantitatively, the final HEP in ASEP is the sum of the diagnosis and execution HEPs. Under the assumption of successful diagnosis, the final HEP is only determined by the probability of making an error in executing post-diagnosis actions, and thus can be optimistically estimated. Although it is difficult to estimate what the true HEPs are given the limited data, the optimism in the HEPs of the SGTR scenarios by Team 1 in the International Study is well illustrated in the HEP pattern (see Figure 1). For the most difficult HFEs 5B1 and 1B, the HEPs are below the lower Bayesian uncertainty bound. In particular, the HEP for HFE 5B1 is 0.025, which shows a large disconnection with the fact the all crews failed that HFE. In addition, the HEPs for HFEs 3B (0.025) and 3A (0.004) appear to be smaller than the actual crew failure rates (2 out of 14 crews failed in HFE 3B and 1 out of 14 crews failed in HFE 3A). However, the optimism did not occur for all the HFEs.

2) Limited guidance for estimating time requirements

The above trend toward optimism in Team 1's HEPs for SGTR scenarios in the International Study is interesting in that ASEP claims to provide generally conservative HEP values. However, where diagnosis was addressed, the HEPs for the LOFW scenarios in the International Study do seem to suggest conservatism. In particular, the HEP for HFE 2B (0.312) not only is above the upper Bayesian uncertainty bound, but also seems to be more conservative than appropriate given a zero crew failure rate. In this case, the main contributor to the conservatism seems to be analysts' conservative assumptions about procedural paths and the allowable diagnosis time in conjunction with the use of the ASEP TRC. It appears that more guidance on estimating time requirements and considering factors that could influence time requirements (e.g., concurrent activities) would strengthen the method.

3) Limited guidance to examine low level cognitive activities

Having learned the lesson from the International Study, the teams in the U.S. Study considered diagnosis in their analyses, and that seemed to help the teams' qualitative analyses. However, the U.S. Study has revealed that even if the analysts decide to explicitly address diagnosis in their analyses, there still is a chance for the analysts to overlook difficulties in cognitive activities if the method guidance is strictly followed. This is caused by the focus of ASEP on procedural steps at a high level (e.g., identification of the initiating event and entry into the appropriate EOP), rather than the diagnosis and cognitive activities involved in following and responding to the steps in the EOPs. That is, lower level cognitive activities, such as interpreting the plant status in the context of the step by step procedures and associated time-limiting conditions need more attention than given in evaluating post-diagnosis tasks. As a consequence, HRA predictions are likely to be limited to the crew's interaction with the main procedural steps and lead to optimistic HRA results by ignoring the difficulties operators would face at the sub-step level.

4) Limited set of performance shaping factors (PSFs)

One primary purpose of qualitative analysis is to understand fully all possible sources of error and the underlying PSFs that impact the reliability of human performance. It could be argued that the analyses in the International Study might have been improved if the HRA teams had explicitly addressed diagnosis. However, even if diagnosis is explicitly included, the method still shows an inability to guide analysts to examine an adequate set of factors that could influence crew behavior for all circumstances. For example, the guidance to evaluate diagnosis/cognitive activities is minimal, and the method relies heavily on its diagnosis TRC with adjustment for only a few PSFs.

5) Limited guidance for choosing PSF levels

When addressing post-diagnosis actions, whether using ASEP or THERP (as mentioned above, ASEP allows the use of THERP in quantification with respect to post-diagnosis actions), decisions need to be made regarding the specific levels of PSFs relative to a given scenario/HFE must be made (e.g., stress levels and execution complexity). In both studies, differences across analysts were seen in the selection of PSF levels, and the differences led to observable variations in the HEPs for the same HFEs. The guidance on those decisions appears to be limited for some situations, which may explain why the teams' decisions on those factors did not appear to correspond well to the factors and conditions observed from crew performance data.

6) Limited insight for error reduction

In general, ASEP can be considered as a PSF-focused method. As mentioned above, it relies heavily on its diagnosis TRC with a few PSF adjustments to address diagnosis. This approach limits the method's ability to discover cognitive mechanisms and/or contextual factors that would lead to human failures, and thus limits its ability to offer insights for error reduction.

4 CONCLUSIONS

Based on discussion above, it is clear that the qualitative analysis performed to support HRA quantification is an important contributor to the adequacy of HRA predictions. This was particularly demonstrated when the method applications did not address the cognitive aspects of performance in implementing procedures even though the initial diagnosis had been completed.

The studies have shown that ASEP, when performed well, can provide good HRA predictions, but variability across analysts can occur, largely due to analysts' decisions about how to apply various aspects of the method. As seen in the studies, analysts are often called upon to make decisions in their analyses, and the guidance of the method is not sufficient or specific enough, so that analysts may have to, more or less, rely on their subjective judgment in interpretation of the guidance.

The differences between the analyst teams observed in the studies underscore the need to enhance the guidance for the application of ASEP. Furthermore, it suggests that piloting of the method (and of this guidance) in view of analyst-to-analyst reproducibility would be warranted.

REFERENCES

Bye A., E. Lois, V. N. Dang, G. Parry, J. Forester, S. Massaiu, R. Boring, P. Ø. Braarud, H. Broberg, J. Julius , I. Männistö, and P. Nelson. 2011. *International HRA Empirical Study—Phase 2 Report: Results from Comparing HRA Method Predictions to Simulator Data from SGTR Scenarios.* NUREG/IA-0216, Vol. 2. US Nuclear Regulatory Commission, Washington, DC.

Bye A., V. N. Dang, J. Forester, M. Hildebrandt, J. Marble, H. Liao, and E. Lois. 2012. Overview and First Results of the US Empirical HRA Study. Proceedings of the 11th International Probabilistic Safety Assessment and Management Conference, Helsinki, Finland.

Dang, V. N., J. Forester, R. Boring, H. Broberg, S. Massaiu, J. Julius, I. Männistö, P. Nelson, E. Lois, and A. Bye. 2011. *International HRA Empirical Study—Phase 3 Report: Results from Comparing HRA Method Predictions to Simulator Data on LOFW Scenarios.* HWR-951, OECD Halden Reactor Project, Halden, Norway. To be issued as NUREG/IA-0216, Vol. 3.

Forester, J., V. N. Dang, A. Bye, R. Boring, H. Liao, and E. Lois. 2012. Conclusions on Human Reliability Analysis (HRA) Methods from the International HRA Empirical Study. Proceedings of the 11th International Probabilistic Safety Assessment and Management Conference, Helsinki, Finland.

Lois, E., V. N. Dang, J. Forester, H. Broberg, S. Massaiu, M. Hildebrandt, P. Ø. Braarud, G. Parry, J. Julius, R. Boring, I. Männistö, and A. Bye. 2009. *International HRA Empirical Study—Phase 1 Report: Description of Overall Approach and Pilot Phase Results from Comparing HRA Methods to Simulator Data.* NUREG/IA-0216, Vol. 1. US Nuclear Regulatory Commission, Washington, DC.

Marble, J., H. Liao, J. Forester, A. Bye, V. N. Dang, M. Presley, and E. Lois. 2012. Results and Insights Derived from the Intra-Method Comparisons of the US HRA Empirical Study. Proceedings of the 11th International Probabilistic Safety Assessment and Management Conference, Helsinki, Finland.

Swain, A. D., and H. E. Guttman. 1983. *Handbook of human reliability analysis with emphasis on nuclear power plant applications.* NUREG/CR-1278-F, U.S. Nuclear Regulatory Commission.

Swain, A.D. (1987). *Accident Sequence Evaluation Program Human Reliability Analysis Procedure.* NUREG/CR-4772/SAND86-1996, Sandia National Laboratories for the U.S. Nuclear Regulatory Commission, Washington, DC, February 1987.

CHAPTER 17

Mirror Neuron Based Alerts for Control Flight Into Terrain Avoidance

Causse, M., Phan, J., Ségonzac, T., Dehais, F.
Institut supérieur de l'aéronautique et de l'espace (ISAE)
Toulouse, France
mickael.causse@isae.fr

ABSTRACT

Controlled flight into terrain (CFIT) accidents occur when an aircraft, under the control of the crew, is flown into terrain (or water) with no prior awareness from the part of the crew of the imminent catastrophe (Wiener, 1977). In commercial aviation, CFIT are among the deadliest accidents but the situation has continuously improved this last decade. In particular, a spectacular fall in the number of fatalities was made possible by the introduction of enhanced ground proximity warning systems (EGPWS). However, CFIT accidents remain the second leading cause of on-board fatalities and several crashes of this category involving airplanes equipped with EGPWS occurred since 2007. The human factor plays a major role in that type of disaster and studies show that visual and auditory alarms are not always taken into account. Yet, when a 'PULL UP' alert is triggered, the pilot has only a few seconds to react in order to avoid the impending CFIT. Most of the time, the procedure is quite simple: the pilot must pull full back on the stick and apply maximum thrust to gain altitude. In this study, we introduced a new type of visual alert specifically dedicated to activate the mirror neurons that appear to play a key role in both action understanding and imitation (Rizzolatti, 2004). Such motor neurons are known to fire either when a person acts or when a person observes the same action performed by another one. We hypothesized that an immediate understanding of a required behavior, displayed by a video that shows the appropriate actions to perform, will activate the mirror neurons and provoke an extremely rapid reaction from the pilots to prevent a potential collision. We designed short videos displayed in the primary flight displays in which virtual avatars explicitly performed the actions on the levers and on the stick. Three pilots completed 10 different flight scenarios during the approach phase with a full motion A320 flight simulator. In some of the scenarios, an alarm was triggered just before an imminent collision and the pilots had to immediately perform a go-around. The results showed that the videos with avatars allowed much shorter reaction times than the regular textual 'PULL UP' alerts. While the anti-collision maneuver was initiated in 7.60 s (SD = 1.83) with the regular alert, video mean reaction time was 1.27 s (SD = 0.31). This encouraging preliminary outcome opens new perspectives on mirror neuron based human machine interfaces.

Keywords: neuroergonomics, mirror neurons, HMI (human machine interface), CFIT (controlled flight into terrain) avoidance

1 INTRODUCTION

Controlled flight into terrain (CFIT) accidents occur when an aircraft, under the control of the crew, is flown into terrain (or water) with no prior awareness from the part of the crew of the imminent catastrophe (Wiener, 1977). In commercial aviation, CFIT are among the deadliest accidents but the situation has continuously improved this last decade: whereas 2152 people died in CFIT accidents during the 1992-2001 period, this number dropped to 1007 people during the 1999-2010 years (Boeing, 2002; Boeing, 2011). In particular, this spectacular fall in the number of fatalities was made possible by the introduction of ground proximity warning systems (GPWS) and enhanced ground proximity warning systems (EGPWS). These systems not only advice aurally the crew (e.g. repetitive 'PULL UP') but they also display a 2D representation of the terrain on a dedicated screen. However, CFIT accidents remain the second leading cause of on-board fatalities and several crashes of this category involving airplanes equipped with EGPWS occurred since 2007, including the Polish president's flight crash. The human factor plays a major role in that type of disaster as accidents analyses reveal that the aircrews do not initiate the go around maneuver because they fail to notice the visual and auditory EGPWS alerts. Yet, when a 'PULL UP' alarm is triggered, the pilot has only a few seconds to react in order to avoid the impending CFIT. Most of the time, the procedure is quite simple: the pilot must pull full back on the stick and apply maximum thrust to gain altitude. In the next two sections we examine the stress hypothesis and the more recent inattentional deafness hypothesis to explain such a visual and especially auditory neglect.

1.1 The stress hypothesis

It may appear surprising that visual and especially auditory alerts can be neglected as these types of alarms are known to present various advantages in emergency situations. They inform the pilots without requiring head/gaze movements (Edworthy, Loxley, & Dennis, 1991) and they provoke faster reaction times than visual stimuli (Wheale, 1981) which allow to be more efficient in emergency situations. However, their aggressive, distracting and disturbing nature (Edworthy, et al., 1991) can considerably increase pilot stress level during warning events, what may provoke a decline of flight performance and decision-making relevance. As a matter of fact, the immediate inclination for pilots can be to find a way to silence the noise, rather than analyzing the meaning of the alert (Peryer, 2005). In 1984, a well known accident (Avianca Flight 011, Boeing 747) demonstrated that an excessive number of auditory alerts may lead pilots to neglect GPWS alerts. From a psychophysiological point of view, a high level of stress is known to provoke a temporary disruption of high level cognitive processes (Porcelli, et al., 2008; Scholz, et al., 2009) and a growing neuroimaging literature demonstrates that this decline of intellectual ability under emotional factor is provoked by the deactivation of prefrontal cerebral structures activity (Qin et al., 2009). There is evidence that emotion may affect the attentional network in a way that attention orienting abilities are impaired (Pecher, Quaireau, Lemercier, & Cellier, 2010). Such impairment of selective attention under arousal seems related to a temporary decline of the activity of the locus coeruleus and a triangular circuit of selective attention (Tracy, Mohamed, Faro, Tiver, Pinus, & Bloomer, 2000). Such an impairment induced by arousal could partially explain the inability to detect visual (Dehais, Causse, Tremblay, 2011) and auditory alerts (Dehais, Tessier, Christophe, Reuzeau, 2009).

1.2 The inattentional deafness hypothesis

Tasks involving high perceptual load consume most of attentional capacity, leaving little or none remaining for processing any task-irrelevant information (Lavie, 1995). Indeed, reduced perceptual processing of task irrelevant information in high-load tasks leads to various forms of inattentional blindness (Mack & Rock, 1998). There is a growing body of evidence for a shared attentional capacity between the modalities of vision and hearing (Brand-D'Abrescia & Lavie, 2008; Santangelo, Olivetti Belardinelli, & Spence, 2007; Sinnett, Costa, & Soto-Faraco, 2006). Given the hypothesized shared attentional capacity between vision and hearing, an engagement in a visual task of high perceptual load is likely to produce a decline of the probability to process a concurrent auditory stimulus. This failure of an auditory stimulus to reach awareness has been recently named inattentional deafness (Koreimann, Strau, & Vitouch, 2009; Macdonald & Lavie). Macdonald & Lavie (2009) showed that up to 79% of participants engaged in a task under high visual load conditions failed to notice a task-irrelevant sound played through headphones. Whereas there are many situations in everyday life in which such phenomenon may be of low importance (eg. the failure to hear someone speaking while engaged in a computer task), inattentional deafness may have important implications with regards to safety, for instance in aviation. Indeed, inattentional deafness may be an additional potential contributive factor to the alarm neglect phenomenon. Numerous displays in modern cockpits are likely to produce this phenomenon and may provoke inattentional deafness, leading pilot to purely fail to notice yet critical alarms.

1.3 Exploiting mirror neuron property to cure alarm negligence

In this study, we introduced a new type of visual alert—which do not require semantic decoding of complex verbal information and do not introduce additional auditory alarm—specifically dedicated to activate the mirror neurons that appear to play a key role in both action understanding and imitation (Rizzolatti, 2004). Such motor neurons are known to fire either when a person acts or when a person observes the same action performed by another one. This type of alert can be a good candidate to inform the pilot of the action to perform, even if this latter is subjected to inattentional deafness or a high deleterious stress. Historically discovered in the rostral part of inferior area 6 (area F5) of the monkey (Rizzolatti, et al., 1990), there is growing evidence that these specialized neurons also exist in human (Rizzolatti, 2005). Functional imaging studies revealed activation in lower part of the precentral gyrus and of the pars opercularis of the inferior frontal during observation of actions made by another individual (Buccino, et al., 2001). The opercularis of the inferior frontal gyrus (basically corresponding to Broadman area 44) likely corresponds to the area F5 in monkey (Petrides & Pandya, 1994). The authors hypothesized that these regions support a mirror system dedicated to action observation/execution matching processes. More recently, a fMRI study of Chong et al. (2008) showed that the right inferior parietal lobe responds independently to specific actions regardless of whether they are observed or executed. Furthermore, magnetoencephalography (Hari et al. 1998) and EEG (Cochin et al. 1999) experiments revealed activation of motor cortex during observation of finger movement. More recently, an EEG experiment of Muthukumaraswamy (2004) showed suppression in the 8–13 Hz (mu) frequency band during the passive observation of object grip. Gastaut et al. (1954) showed that, at rest, sensorimotor neurons spontaneously fire in synchrony leading to large amplitude EEG oscillations in the mu frequency band. In addition, Gastaut et al. (1954) reported desynchronization of these rhythms—thereby decreasing the power of the mu-band EEG oscillations—not only when a subject performed an action, but also while the subjects observed an action executed by someone else. According to Muthukumaraswamy (2004), the mu reduction during the observation of an object grip movement indicates the existence of a brain structure that is functionally comparable to the monkey mirror neuron system. The activation of this hypothesized frontal mirror region in the human brain has also been observed using different modalities, for instance during the observation of static pictures (Johnson-Frey, et al., 2003) or robotic actions (Gazzola, Rizzolatti, Wicker, & Keysers, 2007; Oberman, McCleery, Ramachandran, & Pineda, 2007).

We hypothesized that an immediate understanding of a required behavior, displayed by a video that shows the appropriate actions to perform, will activate the mirror neurons and provoke an extremely rapid reaction from the pilots in order to prevent a potential collision. To test our hypothesis, we designed short videos displayed in the primary flight display (PFD) in which virtual avatars explicitly performed the actions on the levers and on the stick.

2 METHODS

2.1 Participants

Three low experienced male pilots rated for visual flight conditions were recruited from the local flying club. Mean flying experience was 53.33 hours (SD = 32.14). All participants were informed about the GPWS and the associated 'PULL UP' red textual message displayed in the PFD. Each participant provided written informed consent and received complete information on the study's goal.

2.2 Flight scenario

All experiments were conducted in a 3 axis motion A320 flight simulator. The flight scenario was designed with flight instructors to reach a satisfying level of difficulty and realism. During the experiment, each pilot completed 10 identical landing scenarios during the approach/landing phases. Landing occurred in bad meteorological condition (strong crosswind, very low visibility and rain) to increase the stress level of the pilots. In addition, in order to increase the attentional load, the pilots had to count the number of occurrence of a red dot appearing on the screen in front of the pilot monitoring (see Figure 1).

Before the experiment, the pilots were informed that they were in charge of all the decisions and that a go-around procedure might be required in some of the scenarios. The flight scenario started at 2500 feet and the pilot were instructed that they had to maintain a 130 knots speed while piloting the aircraft in instrument flight rule condition with the instrument landing system.

Figure 1 View of the cockpit and the various EFIS. Alerts were displayed in the PFD. The red dot was displayed in peripheral vision on the pilot monitoring PFD screen. ND = Navigation Display; ECAM = Electronic Centralized Aircraft Monitor

2.3 Alerts

During landing 3 and 10, an alert was triggered (between 500 and 600 feet) to notify of an imminent collision. In response, the pilots had to perform immediately a go-around maneuver. All participants received one time the two types of alerts (see Figure 2), the classical TAWS (terrain awareness and warning system) 'PULL UP' and the mirror neuron based alert. The order of their occurrence during landing 3 and 10 was randomized across participants.

Classical TAWS 'PULL UP' alert: Similarly to current classical TAWS alert, the 'PULL UP' red text was displayed in the artificial horizon. In current aircrafts, if the barometric sink rate becomes too severe or if the aircraft is threatened by a terrain hazard, the GPWS voice annunciation "Whoop, Whoop, PULL UP" sounds; the master caution/warning lights illuminate, and the message 'PULL UP' is displayed in red on both PFDs. In our experiment, we exactly reproduced this sequence except that the voice annunciation was removed. This allowed us to focus on the visual effects of the alert and to artificially recreate the inattentional deafness phenomenon.

Mirror neuron based alert: the 212*170 pixels videos were displayed below the airspeed instrument to keep the t-basic visible (airspeed, altimeter, artificial horizon). In addition, as the airspeed in one of the most critical information during a go-around (a minimum speed must be maintained during this procedure), the proximity between the video and this instrument allows to reduce the distance of the ocular saccades.

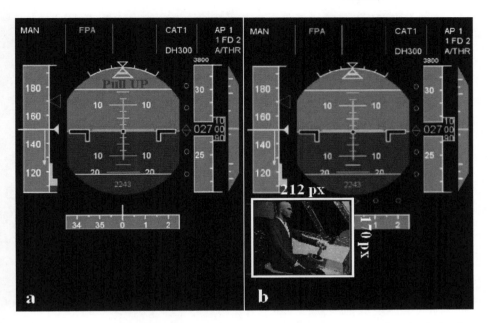

Figure 2 Illustration of the two types of alerts used during the experiment. a: the classical red TAWS 'PULL UP' message; b: the mirror neuron based alert

To assess the efficiency of both alerts, we compared the time taken by the pilots to initiate the go-around action for the regular 'PULL UP' textual message and for the videos. This reaction time corresponded to the time interval between the display of the alert and the time where the stick was set in back position by the pilot to gain altitude (See Figure 3).

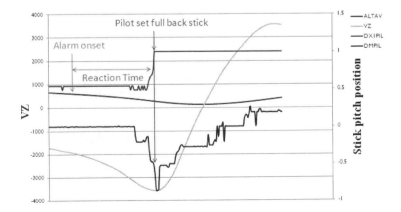

Figure 3 Typical pattern of values after the alarm occurrence. Altav = altitude; Vz = vertical speed; Dx1pil = throttle position; Dmpil = stick pitch position.

3 RESULTS

Given the small size of our sample, we solely present descriptive preliminary results and no statistical tests were performed. All participants reported that they perceived both alerts and that they perfectly understood their meaning and the action that had to be performed. The analysis of the reaction times showed that the avatar videos elicited much shorter reaction times than the regular textual 'PULL UP' TAWS alerts. While the anti-collision maneuver was initiated in 7.60 s (SD = 1.83) with the regular alert, video mean reaction time was 1.27 s (SD = 0.31), see Figure 4.

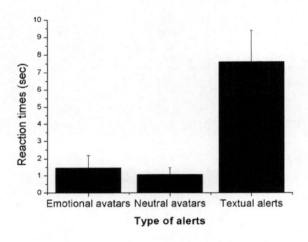

Figure 4 Mean reaction times across the 3 types of alerts. Emotional avatars and neutral avatars allowed faster reactions than classical 'PULL UP' TAWS alerts

4 DISCUSSION

In this experiment we assessed the efficiency of a new type of visual alert in eliciting a very fast reaction from the pilots, namely the pull up maneuver. To purely assess visual aspects and to artificially recreate inattentional deafness, the aural component of the alarm was removed. The results showed that the avatar videos allowed much shorter reaction times than the regular textual 'PULL UP' alerts. This encouraging preliminary outcome opens new perspectives on mirror neuron based human machine interfaces. A future experiment with a larger sample and professional pilots will be conducted to get more conclusive results on the superiority of this type of videos in comparison to the classical TAWS 'PULL UP' alert. In addition, a complementary EEG research will also be conducted and this study will allow to assess the efficiency of these alerts in stimulating mirror neurons. Indeed, the display of stimulus that generates an activation of the mirror neurons is known to provoke a decreased power of the mu-band EEG oscillations

(Muthukumaraswamy, et al., 2004; Oberman, et al., 2007). The observation of such an electrophysiological phenomenon would support that our avatars stimulate these neurons and it would provide evidence on their efficiency to trigger a rapid reaction by imitation in the pilot, even in much degraded situation (workload, stress...) where high level cognitive processes can be strongly altered.

ACKNOWLEDGMENTS

The authors would like to acknowledge Patrice Labedan and Guillaume Garrouste for their precious technical assistance with the experimental set-up.

REFERENCES

Boeing Commercial Airplane Group. *Statistical Summary of Commercial Jet Airplane Accidents: Worldwide Operations, 1959-2001.* . (2002).
Boeing Commercial Airplane Group. *Statistical Summary of Commercial Jet Airplane Accidents: Worldwide Operations, 1959-2010.* Online at: http://www.boeing.com/news/techissues/pdf/statsum.pdf. (2011).
Brand-D'Abrescia, M., & Lavie, N. (2008). Task coordination between and within sensory modalities: Effects on distraction. *Attention, Perception, & Psychophysics, 70*(3), 508-515.
Buccino, G., Binkofski, F., Fink, G. R., Fadiga, L., Fogassi, L., Gallese, V., et al. (2001). Action observation activates premotor and parietal areas in a somatotopic manner: an fMRI study. *European Journal of Neuroscience, 13*(2), 400-404.
Chong, T. T. J., Cunnington, R., Williams, M. A., Kanwisher, N., & Mattingley, J. B. (2008). fMRI adaptation reveals mirror neurons in human inferior parietal cortex. *Current biology, 18*(20), 1576-1580.
Edworthy, J., Loxley, S., & Dennis, I. (1991). Improving auditory warning design: Relationship between warning sound parameters and perceived urgency. *Human Factors: The Journal of the Human Factors and Ergonomics Society, 33*(2), 205-231.
Gastaut, H. J., & Bert, J. (1954). EEG changes during cinematographic presentation (Moving picture activation of the EEG). *Electroencephalography and Clinical Neurophysiology, 6*, 433-444.
Gazzola, V., Rizzolatti, G., Wicker, B., & Keysers, C. (2007). The anthropomorphic brain: the mirror neuron system responds to human and robotic actions. *Neuroimage, 35*(4), 1674-1684.
Johnson-Frey, S. H., Maloof, F. R., Newman-Norlund, R., Farrer, C., Inati, S., & Grafton, S. T. (2003). Actions or hand-object interactions? Human inferior frontal cortex and action observation. *Neuron, 39*(6), 1053-1058.
Koreimann, S., Strau, S., & Vitouch, O. (2009). *Inattentional deafness under dynamic musical conditions.* Paper presented at the 7th Triennial Conference of European Society for the Cognitive Sciences of Music (ESCOM 2009)), Jyväskylä, Finland.
Lavie, N. (1995). Perceptual load as a necessary condition for selective attention. *Journal of Experimental Psychology: Human Perception and Performance, 21*(3), 451.
Macdonald, J. S. P., & Lavie, N. Visual perceptual load induces inattentional deafness. *Attention, Perception, & Psychophysics*, 1-10.
Mack, A., & Rock, I. (1998). *Inattentional blindness*: The MIT Press.

Muthukumaraswamy, S. D., Johnson, B. W., & McNair, N. A. (2004). Mu rhythm modulation during observation of an object-directed grasp. *Cognitive Brain Research, 19*(2), 195-201.

Oberman, L. M., McCleery, J. P., Ramachandran, V. S., & Pineda, J. A. (2007). EEG evidence for mirror neuron activity during the observation of human and robot actions: Toward an analysis of the human qualities of interactive robots. *Neurocomputing, 70*(13-15), 2194-2203.

Pecher, C., Quaireau, C., Lemercier, C., & Cellier, J. M. (2010). The effects of inattention on selective attention: How sadness and ruminations alter attention functions evaluated with the Attention Network Test. *European Review of Applied Psychology*.

Peryer, G., Noyes, J., Pleydell-Pearce, K., & Lieven, N. (2005). Auditory alert characteristics: a survey of pilot views. *International Journal of Aviation Psychology, 15*(3), 233-250.

Petrides, M., & Pandya, D. N. (1994). Comparative architectonic analysis of the human and the macaque frontal cortex. *Handbook of neuropsychology, 9*, 17-58.

Porcelli, A., Cruz, D., Wenberg, K., Patterson, M., Biswal, B., & Rypma, B. (2008). The effects of acute stress on human prefrontal working memory systems. *Physiology & Behavior, 95*(3), 282-289.

Rizzolatti, G., & Craighero, L. (2004). The mirror-neuron system. *Annu. Rev. Neurosci., 27*, 169-192.

Rizzolatti, G. (2005). The mirror neuron system and its function in humans. *Anatomy and Embryology, 210*(5), 419-421.

Rizzolatti, G., Gentilucci, M., Camarda, R., Gallese, V., Luppino, G., Matelli, M., et al. (1990). Neurons related to reaching-grasping arm movements in the rostral part of area 6 (area 6aβ). *Experimental Brain Research, 82*(2), 337-350.

Santangelo, V., Olivetti Belardinelli, M., & Spence, C. (2007). The suppression of reflexive visual and auditory orienting when attention is otherwise engaged. *Journal of Experimental Psychology: Human Perception and Performance, 33*(1), 137.

Scholz, U., La Marca, R., Nater, U., Aberle, I., Ehlert, U., Hornung, R., et al. (2009). Go no-go performance under psychosocial stress: Beneficial effects of implementation intentions. *Neurobiology of Learning and Memory, 91*(1), 89-92.

Sinnett, S., Costa, A., & Soto-Faraco, S. (2006). Manipulating inattentional blindness within and across sensory modalities. *The quarterly journal of experimental psychology, 59*(8), 1425-1442.

Tracy, J. I., Mohamed, F., Faro, S., Tiver, R., Pinus, A., Bloomer, C., et al. (2000). The effect of autonomic arousal on attentional focus. *Neuroreport, 11*(18), 4037.

Wheale, J. L. (1981). The speed of response to synthesized voice messages. *British Journal of Audiology, 15*(3), 205-212.

Wiener, E. L. (1977). Controlled flight into terrain accidents: System-induced errors. *Human Factors: The Journal of the Human Factors and Ergonomics Society, 19*(2), 171-181.

CHAPTER 18

Computer Technology at the Workplace and Errors Analysis

I. Bedny, W. Karwowski, G. Bedny

Evolute.
Louisville, KY, USA
innayarosh@gmail.com
University of Central Florida,
FL, USA
wkar@ucf.edu
Evolute
Lousville, KY, USA
gbedny@optonline.net

ABSTRACT

This paper presents a new approach to human errors analysis that can be utilized in man-machine and computer based systems. Systemic-Structural Activity Theory (SSAT) is the basis for this analysis. Principle of errors classification is considered from individual-psychological (sub-system level) and business organization (system level) prospective.

Keywords: human-computer systems, self-regulation, actions, classification of errors

1   INTRODUCTION

Error analysis is a critical factor in evaluation of precision and reliability of human performance. It has been estimated that human errors are the main cause of accidents and incidents in complex human-machine systems. Presently there is a significant trend toward computerization of human-machine systems and therefore analysis of human errors when performing computer-based tasks becomes very important. We can outline several major factors which effect human errors in HCI system: quality of developed software, user qualifications (novice versus highly trained user), individual features of a user, reliability of technical components of computerized system, protection of computer system against viruses, etc.

The reliability of computer systems is extremely important for small and mid size businesses that keep their financial and client pull information in the database. Computer systems' hardware should be kept up to date; the operating systems have to be ungraded and patched according to the vendor dictated schedule. The database has to be backed up regularly and the backup media should be kept in the safe place different from the computer system location. In case of any disaster the system then can be restored from the backup.

In complex and potentially hazardous computerized systems the role of human errors significantly increased. In such systems requirements for reliable human performance should be significantly increased and at the same time vulnerability to human errors decreased. Another aspect of such system is that consequence of human errors not always become immediately apparent, and what is the cause of system malfunctioning is not clear. This work presents in abbreviated manner some aspects of human error analysis in HCI system from systemic-structural activity theory (SSAT) perspective.

2 ERRORS TAXONOMY: AN INDIVIDUAL-PSYCHOLOGICAL APPROACH

We distinguish two major groups of human errors in human-computer systems. One group of errors is considered on business organizational level (systemic level) and another one is on individual-psychological level (sub-system level). One of the important issues of error analysis on the system level is the relation between formal structure of business organization and users' opportunity to perform their individual duty. Individual-psychological level can provide an error analysis during performance of a specific task. A starting point in understanding of operator errors at the individual-psychological level is the development of various error taxonomies, which can be utilized during task analysis.

Any taxonomy can be helpful if it can be utilize for the error analysis of human performance. In cognitive psychology the categories of human errors are derived from information processing approach. In SSAT classification of errors is derived from the analysis of activity self-regulation and classification of human actions. Below we present error taxonomy, which has derived from SSAT. This taxonomy includes five basic criteria and their parameters and dimensions and it corresponds to the hierarchical decomposition of criteria for error analysis (Table 1).

Table 1 CRITERIA FOR ERROR ANALYSIS

Criteria	Parameter	Dimension
General characteristics of the errors at the system level	In what particular system or subsystem errors occurred and the time of their appearance	At what stage of system operation errors occurs: during normal operation of the system; during testing and evaluation of the system function; during unexpected complication of the system functioning.
	Working conditions of the system or subsystem in which errors occur	Good, bad, overloaded, underloaded, etc.
	External manifestation (errors consequences)	Consequential, non-consequential
General characteristics of the errors at the task analysis level	Relation of errors to the task in which they occur.	In what task do errors occur
	Detectability	Obvious, hidden
	Operator's awareness of the errors	Operator is aware, unaware about the error.
	Existing causes	Predictability
	Probabilistic characteristics: Typical Expectedness Kind Quantitative	Usual, unusual Expected, unexpected Constant, variable Frequency of occurrence and number of occurrences
	Functional characteristics of errors.	Procedural or reproductive, creative, perceptual, cognitive, behavioral, symbolic, calculative.
Position of the errors in the structure of the operator's activity	Goal formation and evaluation of the meaning of the situation and task.	Incorrect understanding and formulation of goals, conceptual and dynamic mental model.
	Evaluation of strategies of task performance.	Wrong and/or untimely selected strategies of activity, or incorrect

		transition from one strategy to another. Relationship between strategies of task performance and errors. Analysis of objective and subjective criteria of success, existing feedback, etc.
	Evaluation of specific characteristics of tasks: Relationship between complexity and difficulty.	Under or overestimation of task difficulty, self-evaluation of personal abilities and task requirements.
	Sense of task and motivation. Personal attitudes.	Personal significance of task and level of motivation to follow safety requirements.
	To what particular cognitive actions do errors belongs.	Errors resulting from sensory and perceptual actions, mnemonic actions, thinking actions, decision-making actions.
	To what particular verbal and motor actions do errors belong	Errors connected with discrete and continuous motor actions, verbal errors, or errors connected with undesired involuntary responses.
	Inadequacy of activity regulation level.	1) Level of stereotypy or automaticity of performance; 2) Level of the conscious regulation of activity in terms of acquired rules and familiar strategies; 3) Level of regulation of activity based on general knowledge, principles, and heuristics.
	Team performance strategies.	Errors resulting from social-organizational aspects of task performance.

Cause of errors	Errors stem from idiosyncratic (personal) characteristics, equipment design or organizational characteristics of the system, interface characteristic	Errors caused by human factors (human erroneous actions were discovered), or errors are derived from technical factors of the system.
	Errors stem from functional state of users or operators.	Errors caused by fatigue, boredom, monotony, decreased vigilance.
	Stress-producing factors.	Time limitations, danger and other external influences.
	Technical factors.	Incorrect distribution of functions between human and machine, inadequate information and instructions, inadequate design of equipment.
	Organizational factors.	Interference of supervision, inadequate work/rest schedules, excessive or insufficient information.
	Operator's experience.	Insufficient training and experience, insufficient knowledge and skills, inability to select required strategy of activity, and timely transfer from one strategy to another.
	idiosyncratic (personal) features	Operator is unsuited for work because of inadequate intelligence, features of attention and physical fitness or emotional stability.
Analysis of consequences of errors from three points of view.	Influence on efficiency of the system.	System malfunctioning or shutdown, failure to achieve goal in assigned time, accident.
	Influence on operator's activity.	Incorrect cognitive or

		behavioral actions, incorrect sequence of actions, untimed performance of actions or their farther performance.
	Influence on operator's state.	Produce stress, fatigue, loss of attention, leads to inability to continue activity, etc.

The proposed classification of errors integrates a large number of various factors, accumulated in the activity theory and cognitive psychology into a unified system and offers a method of psychological analysis of the causes of errors.

As an example, let's consider a possibility to use error analysis based on activity theory. The cause of erroneous actions may be the incorrect formation of the goal of task or actions, incorrect interpretation of the goal, inadequate selection of strategies of goal attainment. For example, the operator incorrectly formulates the goal of actions and as a result performs wrong actions. In SSAT this is error at the goal formation stage. In cognitive psychology according to Norman (1988) this will be mistake. But when the operator correctly formulates the goal of actions but accidently performs the wrong action, according to activity theory and cognitive psychology this is a slip. Let us consider another example. Errors more often can occur when the level of activity self-regulation is inadequate. For example, in face of unpredictable changes of situation the level of stereotypy or automaticity of activity regulation during task performance may result in errors. In general, the conclusion about stems of errors from personal factor could me made only when the erroneous human actions are discovered. Hence the concept of goal and action, as they are understood in activity theory, help us to describe and classify errors and develop methods of their prevention. Errors during task performance depend on user's strategies, adequacy of mental model of task, evaluation of task difficulty, significance, subjective criteria of success, and feedback (Bedny, Karwowski, 2007; Sengupta, T., Bedny, I. S., & Karwowski, W., 2008; Bedny, Karwowski, eds. 2011).

According to Landa (1976), Zarakovsky and Medvedev (1979), and others we can extract a threefold level of regulation of activity: a) the level of stereotypy or automaticity of performance; b) the level of conscious regulation of activity in terms of acquired rules and familiar strategies; c) the level of regulation of activity based on general knowledge, principles, and heuristic strategies. All of these levels have a hierarchical organization. We can see from the table that inadequacy of activity regulation level and conditions of performance may be the reasons for errors. For example, in the face of unpredictable changes of situation, stereotyped methods of performance may result in errors. In dynamic situations with low levels of predictability sequence of events, the ability to use flexible strategies of activity assumes greater significance. In contemporary working conditions and in computer-based tasks in particular, errors occur more often in difficult problem-solving situations and therefore concepts of strategy, self-regulation etc. becomes particularly important. Errors may vary widely depending on the evaluation of task difficulty and its significance, selection of subjective criteria of success, etc. Precision of performance and produced errors should be considered as an aspect of self-regulation and derived from them strategies of performance. For example it was discovered that accuracy with which a pilot can read an aviation instrument often depends more on the significance (subjective importance) of this instrument than on visual features of the instrument.

3 EXAMPLE OF ERROR ANALYSIS IN HCI AT THE BUSINESS ORGANIZATIONAL LEVEL

Presented above Table 1 can be utilized for the error analysis at the individual-psychological level. Let us consider in an abbreviated manner error analysis at the business organization level (systemic level) related to utilization of computers. Below we present an example of possible causes of errors at system level.

1) Errors in software design and coding: a) misinterpretation of requirements; b) poor testing; c) poor or no documentation; d) poor quality of user interface.

2) Errors caused by the users or by the organization that utilize the computer system: a) organization provided wrong specifications for the computer system;
   b) key entry errors during data population process.

3) Errors that emerge during the system implementation or upgrade. This kind of errors is usually a result of either miscalculation in technical requirements or lack of adequately trained personnel: a) hardware/operating system failure during the implementation; b) poor system architectural design; c) incompatibility of hardware/operating system/data base with the business needs or inadequate training of staff to use the newly implemented system.

A separate group of system level errors is related to the integration of various components of computer system and coordination of functioning of such components that we will call *systems consolidation errors*. Mergers and acquisitions are very popular these days. These processes lead to consolidation and integration of computer systems and databases of the companies involved in such process. If the information has to flow from one of the consolidated company database to the other company database (DB) the tables in both databases should have identical structure. The common error is to update one of such DBs to a new version or to change its structure without making the same changes in the DB that shares information real time with the first one.

Let's consider an example of systems consolidation errors. One company uses the system for Human Resources Department. It bought a company that uses the same system for both Human Resources and Payroll Departments. Payroll system has to be updated every two month because of the tax updates. The Human Resources system does not have to be updated that frequently. After the acquisition both companies shared their information. The Payroll system had to be updated. Every such update includes changes in the software and in the structure of the DB.

It has been decided not to update the Human Resources system of the first company to the same version.

Figure 1 depicts the structure of the Payroll Deduction table with the names of its columns, the size of the columns and their description. This figure shows the table structure and the characteristics of its columns.

Figure 2 depicts the fragment of the Payroll Deduction table with the data stored in this table. Each column in this table holds data for various types of deductions. Columns in this figure correspond to the rows on Figure 1.1. We choose this table as an example because update of the Payroll system included structural changes of this table and as a result the Human Resources system was unable to access this table. During the update of the Payroll system a new column with tax related information has been added to the Payroll Deduction table. This change in the table structure made it different from the identical table in the Human Resources system, which lead to inability of these two systems to share information. This is an example of systems consolidation errors. Taxation changes don't affect Human Resources system. So, it has been decided to leave it out of the update. However, these two systems share information real time and their table should be in sink. Otherwise the information flow is disrupted.

The above described error should be related to system integration and/or consolidation errors and needs a system level analysis.

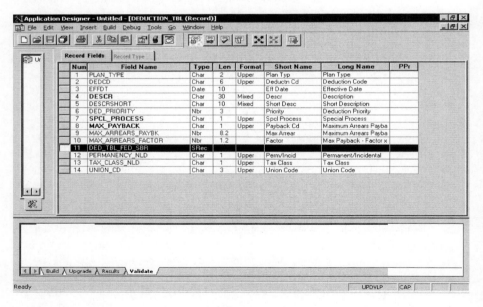

Figure 1 Structure of the Payroll Deduction Table

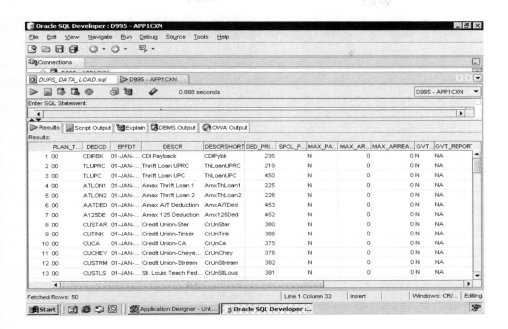

Figure 2 Fragment of Payroll Deduction Table.

Another kind of errors that arises during the systems integration/consolidation is related to the errors of data load and data population. When computer systems are consolidated it becomes necessary to merge massive DBs together. Such a process is usually facilitated by writing special software and testing it in especially dedicated test DBs that have the same structure as the real ones. The typical error in such testing is to check only newly created DB. It's very important to compare the new DB with the sources of it population to make sure that the information has not been lost or corrupted. Consolidation of the systems always introduces additional risk of errors and failures. The businesses usually concentrate on reducing the cost of such consolidation.

CONCLUSIONS

According to the statistical data, 60% of small and mid size businesses fail within the next two-month after losing their data base. It's very important for the reliability of the company's computer system that the employee who is responsible for its maintenance is familiar with the computer terminology and can follow the upgrade and backup instructions.

It should be taken into consideration that the low computer user comfort level might negatively affect the functioning and reliability of the human-computer systems. In some cases it will result in unwillingness to use computer for certain tasks, in other cases it will lead to human errors or system failures that will result in the decrease of productivity, financial loses, decrease in quality of product, etc.

The current work briefly presents error analysis principles that derive from SSAT. The presented method can be used for error analysis of man-machine and computer based system and can be especially beneficial during consolidation of computer systems during mergers and acquisitions. The relation between the structure of human activity, activity strategy and human errors is analyzed.

REFERENCES

Bedny, G. Z. & Karwowski, W. (2007). A systemic – structural theory of activity.
    Application to human performance and work design. Taylor and Francis.
Bedny, G. Z., Karwowski, W. (Eds.). (2011). Human-computer interaction and operators'
    performance. Optimizing work design with activity theory. Taylor and Francis.
Landa, L. M. (1976). Instructional regulation and control. Cybernetics, algorithmithation and
    heuristic in education. Educational technology publications. Englewood Cliffs, New
    Jersey.
Norman, D. A. (1988). The psychology of everyday things. New York: Harper & Row
Sengupta, T., Bedny, I. S., & Karwowski, W. (2008). Study of computer based tasks
    during skill acquisition process, 2nd International Conference on Applied Ergonomics
    jointly with 11th International Conference on Human aspects of Advanced
    Manufacturing (July14-17, 2008), Las Vegas, Nevada.
Zarakovsky, G. M., Medvedev, V. I. (1979). Classification of operator errors. Technical
    aesthetics, 10, 5-7.

CHAPTER 19

A Study on Document Style of Medical Safety Incident Report for Latent Factors Extraction and Risk-Assessment-based Decision Making

Keita Takeyama, T. Akasaka, T. Fukuda, Y. Okada

Keio University
3-14-1, Hiyoshi, Kohoku, Yokohama 223-8522, Japan
keiokeita@z3.keio.jp

ABSTRACT

In order to prevent future medical troubles, incident cases are collected. From the collected data, the causes/factors are analyzed, and prevention measures are proposed. However, in many medical spots, quality of data obtained from incident report documents is very low. So, analysis of factors becomes insufficient, and many prevention measures cannot defeat latent factors. Therefore, we intend to propose an incident report support system. This system builds on three foundations. First is an incident report form to easily identify factors that lead to accidents. Second is a method to decide priority of an incident by analyzing risk from not only a subjective viewpoint but also an objective viewpoint. Third is a matrix of measures to prevent troubles, as well as to plan measures for even those lacking knowledge or experience and to reduce time and burden in considering measures. We developed such an incident report support system application running on a PC. Users only have to click items they think are relevant. Once items are specified, the application processes them and gives feedback based on the items. 21 out of 23 safety managers in 5-units said that they would want to use this application, which has feedback on evaluating priority of an incident and listing radical measures to prevent troubles. By using an incident report support system, safety activity can expand beyond not only safety managers but also on site medical staffs. With such a support system, staffs can come together and cope with safety activity, which is expected to lead to safety improvements.

Keywords: human factors, human error, medical safety, medical trouble incident

1   INTRODUCTION

For several years, safety activities such as preventing medical accidents have been emphasized in Japanese medical fields. In order to maintain medical safety, it is important to encourage safety activities in daily work. One of the major safety activities is to analyze medical trouble incident cases to prevent future medical accident. This safety activity process involves analyzing factors associated with a medical trouble incident from collected data and then taking preventive measures. Such medical trouble incidents data are collected by submitting reports called "incident reports." However, in many medical spots, the purpose of the incident report is to get a grasp on an occurrence and heighten each person's awareness. But, to prevent future accidents, it is important that various latent factors lead to preventive measures. Performing incident-reporting procedures, such as those shown in Figure 1, can do this.

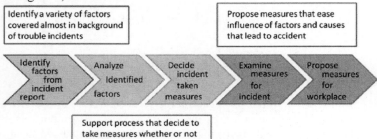

Figure 1 Flow of the study on advanced reporting system

The purpose of this study was to help prevent accidents by using an incident report support system we propose. So, three contents of the system were made. The first content component is an "incident report form," which is used to identify various latent factors in the background of trouble incidents. The second content component is a "method to decide priority of incident," which is used by safety managers to analyze incident risk. The third content component is a "matrix of measures to prevent troubles," which is used by safety managers to take measures associated with latent factors. Below, we included the methods to decide priority of incident and develop a matrix of measures to prevent trouble in the incident report form. The objective is to improve substance of medical safety activity.

When a medical trouble incident case occurs, the person concerned examines the content, the factor, and the measure of the medical trouble incident etc. at an individual level, and then writes up an incident report. A risk manager compiles submitted reports and, in turn, staffs discuss in each unit's conference based on some of the reports. Thus, the basic process is that factors are identified and analyzed from incident reports and preventive measures are executed. The items described in the conventional incident report are contents, place, factors of occurrence, work contents, improvement plan, influence degree, etc. The essential characteristic of the generation content and generation factor processes is that it is a free description form and others are check forms. In this way, medical trouble incident cases are collected, but the content, such as "It was my attention shortage" and "It was my confirmation shortage," is often written in the point of view of the generation factor and generation content because the hospital's staff doesn't understand what they should write. And because the staff easily has a negative image of the report of the failure, most of these reports uncover superficial factors, such as insufficient verification, not noticed, and mistake. And this prevents radical measures to taken in preventing troubles because information included in incident reports are insufficient. So, incident reports are merely a collection of facts. Administrator examines many measures to prevent troubles based on factors. But, all measures can't be performed. Moreover, staffs can't handle measures to prevent troubles at the same time because the administration of the organization concerns various limitations such as expense, manpower, and time. Therefore, it is desirable to take measures estimated as important in safety activity. But the conventional incident report is unsuitable for assessment of factors and decisions to take such measures. And these decisions depend on expert's experience and knowledge. Yet, a formal method of decision-making hasn't been established.

2 PROPOSAL OF THREE ITEM

2.1 Incident Report Form

Most of the information one an get from a conventional incident report involves superficial factors, such as insufficient verification, not noticed, and mistake. This leads to a lack of taking radical measures because information included in conventional reports are insufficient. To realize radical measures, it is important to consider various latent factors. We thus made a new form of incident report to easily identify factors that lead to accidents by identifying various factors that can get information regarding an incident from various aspects. We included not only subjective information but also key information considered factors of the incident.

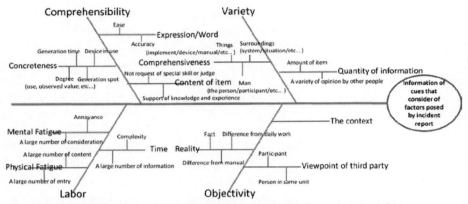

Figure 2 Group of factors related the incident report to write latent factors on check form.

Content of the new form of incident report followed research methods. First, we reinvestigated over 100 medical trouble incident cases in detail, and identified about 40 primary factors. Most all of these primary factors were obtained by interviewing two medical staffs, and were not written in report documents. Based upon the primary factors, secondary factors and latent factors were introduced by Root Causes Analysis. By arrangement of factors, 62 document styles were designed to identify, in detail, various latent factors that aren't identified by conventional methods. Figure 3 is one of the 62 document styles. The new form of incident report includes question items key to noticing latent factors and listed question items that correspond to each of these factors, which are arranged and collected by group (see Figure 2).

Setting factors on it
Lack of knowledge
Lack of experience
Imperfect of estimate
Irrelevant to time schedule
Imperfect content of manual
Interruption of work
Poor concentrating environment
Ambiguous/indefinite direction
Storage of fatigue
Mental stress
Lack of a hand
Frequent moving to out of unit
Poor working environment
Imperfect of record
Inexperience to use machine
Inattention
Working at the same time
Prejudice
Neglecting safety
Decline motivation
Lack of confirmation
Ignoring procedure
Working alone
Lack of awareness
Lack of communication
Lack of explanation for patient
Condition of patient

Figure 3 New form of incident report and setting factors

2.2 Method to Decide Priority of Incident

Not all incidents can have measures because of various limitations. It is desirable for safety activity that measures are taken from collected incident reports for incidents starting with the highest priority and working down in decreasing order. But now, evaluating priority of incident depends on experience or knowingness of safety manager. It is desirable to evaluate priority from not only a subjective viewpoint but also an objective viewpoint. So, this study proposed a method to objectively evaluate the risk of incidents.

Various factors, and in full detail can be identified by the method we proposed. From these identified factors, one can then work to eliminate and reduce high risk factors to, in turn, reduce the possibility of human error occurrences. So by using TESEO (Tecica Empirica Stima Errori Operatori) and HEART (Human Error Assessment and Reduction Technique), which evaluate error probability of incident as references, we proposed a calculation method that evaluates the possibility of human error occurrence for each incident by evaluating factor conditions identified from the new form of incident report. The possibility of human error occurrence is called HEP (human error possibility), and is calculated as:

$K1 = K1_0 \times K1_1 \times K1_2$
$K2 = K2_0$
$K3 = K3_0 \times K3_1 \times K3_2$
$K4 = K4_0 \times K4_1$
$K5 = K5_0 \times K5_1 \times K5_2$
$K6 = K6_0$
$K7 = K7_0 \times K7_1$
$K8 = K8_0 \times K8_1 \times K8_2 \times K8_3 \times K8_4 \times K8_5$
$_n HEP = K1 \times K2 \times K3 \times K4 \times K5 \times K6 \times K7 \times K8$
$_n R = R1 \times R2 \times R3$
$HEP = {_n HEP} \times (1 - {_n R})$

When you calculated K1, you refer 4A & 4B & 4C.
When you calculated K2, you refer 4D.
When you calculated K3, you refer 4E & 4F & 4G.
When you calculated K4, you refer 4H & 4I.
When you calculated K5, you refer 4J & 4K & 4L.
When you calculated K6, you refer 4M.
When you calculated K7, you refer 4N & 4O.
When you calculated K8, you refer 4P & 4Q & 4R & 4S & 4T & 4U.
When you calculated R1, you refer 4V.
When you calculated R4, you refer 4W.
When you calculated R3, you refer 4X.

4A

Work frequency	$K1_0$
This time is first	0.01
Already several times	0.005
A few times by years	0.001
A few times by months	0.0001
A few times by weeks	0.00005
routinely (almost all day)	0.00001

4B

Procedure (Q9)	(Q9-3)	$K1_1$
Had been decided	Did on procedure	1
	Coped with content needed to cope with in the situation	10
	Work changed by instruction of that day existed	20
	Looking situation of the day, I got approval of the boss and changed or left out procedure	5
	By my judge, I changed or left out procedure	100
	I don't know	20
Had not been decided		20

4C

Difficulty (Q23)	$K1_2$
Very difficult	100.
Difficult	50.
A little difficult	10.
Normal	1.
Work reflectively	0.1
I don't know	20.

4D

Behind time schedule (Q1)	$K2_0$
Very behind time	10.
Behind time	2.
A little behind time	1.
Not behind time	0.5
I don't know	1.5

4E

Work experience of years	$K3_0$
Under one year	0.1
From one to two years	0.05
From three to five years	0.01
Over five years	0.001

4F

Health condition (Q4)	$K3_1$
Not good of that day	10
A little not good of that day	5
Feel is a little strange of that day	2
Often not good	3
Always not good	10
No problem	1

4G

Mental condition (Q5)	$K3_2$
Not good of that day	10
A little not good of that day	5
Feel is a little strange of that day	2
Often not good	3
Always not good	10
No problem	1

4H

Poor transmitting information in unit (Q2)	$K4_0$
Had been occurred	2.
Sometimes occurred	1.75
Occurred	1.5
Not occurred	1.
I don't know	1.5

4I

Poor transmitting information yourself (Q3)	$K4_1$
Had been occurred	2.
Sometimes occurred	1.75
Occurred	1.5
Not occurred	1.
I don't know	1.5

4J

Business (Q7)	$K5_0$
Very busy	5
Busy	4
A little busy	2
Only me busy	5
Not busy	1

4K

Did multi-tasking at the same time this time (Q17)	$K5_1$
Over three task	10
Two task at the same time	8
Temporarily multi-tasking	5
Interrupting task make relevant task suspend	3
Not did	1

4L

Distance traveled in this work (Q18)	$K5_2$
Very long	1.5
Long	1.25
Not short	1.
Not feel a load	0.5
Short	0.1
I don't know	0.1

4M

Happened sudden incident (Q8)	$K6_0$
Happened unexpected incident	3.
Happened but had been expected	2.
happened trivial incident	1.5
Happened but ignored	1.25
Not happened	1.
I don't know	1.

4N

Used medical equipment (Q12)	Q12-3	$K7_0$
No		1.
Yes	Had not been used	5.
	Before several years used	4.
	used by year	2.
	Several times used by month	1.5
	almost all day	1.
	I don't know	2.

4O

Shortage of item (Q20)	$K7_1$
Shortage, but did just as it as	3.
Go other unit for item as shortage	2.
Use substitute temporarily as shortage	1.5
Not shortage	1.
I don't know	1.5

4P

Relevant brightness (Q13)	$K8_0$
Not enough on a daily basis and that day	2.
Not enough brightness	1.5
Cannot say that enough and not enough	1.
Enough	0.8
I don't know	1.5

4Q

Relevant temperature and humidity (Q14)	$K8_1$
An uncomfortable feeling	2.
A little uncomfortable feeling	1.5
Not comfort or uncomfortable feeling	1.
Comfort	0.8
I don't know	1.5

4R

Situation induced a risk (Q15)	$K8_2$
Always such situation	2.
Lots of spots such situation	1.75
A part of spot such situation	1.25
No problem	1.
I don't know	1.25

4S

Room of space that work with several people (Q19)	$K8_3$
Not absolutely	2.
Very small	1.75
Small	1.5
Safe if we are careful each other	1.25
Enough room	1.
I don't know	1.25

Figure 4-1 Relation between scoring and the condition of the item in the new incident report

4T

Appropriate place of item (Q21)	$K8_4$
Not appropriate, but I didn't say everybody	2.
Did an oral report as not appropriate	1.75
Made note as not appropriate	1.5
No problem	1.
I don't know	1.25

4U

Cleaned or classified similar item (Q22)	$K8_5$
Not cleaned or classified, but I didn't say everybody	2.
Did an oral report as not cleaned or classified	1.75
Made note as not cleaned or classified	1.5
No problem	1.
I don't know	1.25

4V

Involved staff (Q6)	R1
Always several persons existed	0.99
Always a person existed	0.99
Temporarily existed	0.95
A person existed at work but other person had been existed before work began	0.9
Not existed	0.9
I don't know	0.99

4W

Did confirmation of work (Q11)	R2
Always and this time not did	0.
Always did, but this time not	0.3
Did check with eyes, but not did pointing and calling	0.8
Did check with eyes and pointing and calling	1.
I don't know	0.8

4X

Have a person that is easy to communicate (Q16)	R3
Not always	0.5
Now not	0.6
Not either	0.8
Exist in other unit	0.9
Exist at least one person in same unit	0.95
Exist several persons	1.

Figure 4-2 Relation between scoring and the condition of the item in the new incident report

Based on the score of HEP, the priority of an incident for which measures have been taken is decided and decision making for analysis is supported. Table 1 shows suggestions that are given to safety managers based on the evaluation value of HEP.

Table 1 Priority of measures according to HEP

$\log(HEP \times 10^9)$	Priority	Contents
Over 6	A Generation of trouble not leading to accident	Need for taking measures directly
		Stop work until taking measures
		Need for paying sufficient management resources
3.5 ~ 6	B Not need of treatment	Need for taking measures without delay
		Should stop work until taking measures
		Paying management resources preferentially
Under 3.5	C Not special damage	Taking measures according to need

2.3 Matrix of Measures to Prevent Troubles

Radical measures to prevent troubles are often not taken because such things as attention, scolding, and strengthen awareness are merely listed as requisite actions. For this effort, measures that are considered radical were listed and arranged into a matrix of measures to prevent troubles that were connected with setting factors. By using it as a reference, the matrix of measures to prevent troubles can support the planning of measures for even those lacking knowledge or experience and to reduce time and burden considering measures to prevent troubles. Figure 5 shows how considered measures to prevent troubles were connected with setting factors.

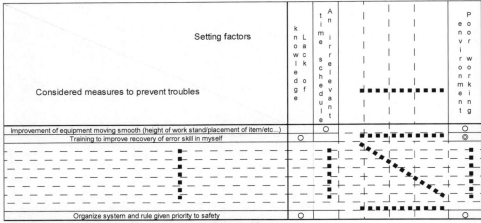

Figure 5 A part of matrix of measures to prevent troubles

3   SUPPORT TO INPUT DATA IN INCIDENT REPORT SYSTEM

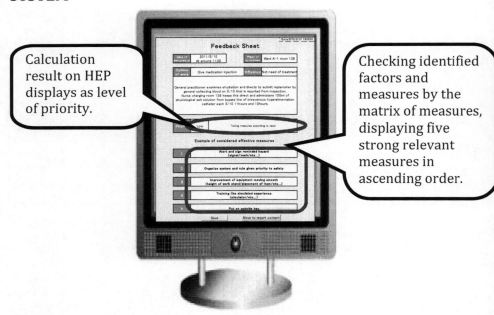

Figure 6 Example of picture in developed system

We developed an application running on a PC. The application provides users with an easy to use user interface. Users only have to click items they think are relevant. Once items are specified, the application processes them and gives feedback based on the items.

The developed system was investigated to assess its subjective validity. In the results of the investigation, 21 out of 23 safety managers in 5 units said that they would want to use the support system, which supports evaluating the priority of incidents and listing radical measures to prevent troubles. Their comments were as follows:
- Although analyzing felt difficult, it is useful for the analyzer because it leads the evaluation and measures easily, and it is easy to give feedback to the staff quickly.
- This system is very worthy as a reference for a person that does not easily notice factors or consider measures when filling out an incident report.
- So I have no confidence in subjective decisions, my decision making for analysis is supported by this system having objective decisions and an index for the answer is got.
- Although similar measures to prevent troubles have been considered, this system is worthy as a reference for considering measures, and with it the possibility of considering new measures increases.
- This system could support identifying various latent factors that have the possibility of leading to an accident, and the burden and difficulty of the staff can be reduced by checking via the input sheet.

To the contrary, 2 safety managers that did not want to use the system had the following comment:
- I can't depend on this system because I don't think that the listed measures to prevent occurrences really related to the incident written by the incident report.

This system proposes an index of analysis and decision-making measures to prevent troubles for safety manager by guiding objective evaluation and providing measures to prevent troubles. Until now, collecting and analyzing incidents depended on safety manager having special knowledge and experience. But by using the incident report support system, safety activity can expand beyond the safety manager to the medical staff on site. Through an improved incident report support system, staffs can come together and cope with safety activity, which is expected to lead to safety improvements.

4 CONCLUSION

In order to prevent future medical troubles, incident cases are collected. From the collected data, causes/factors are analyzed, and prevention measures are proposed. However, in many medical spots, quality of data obtained from incident report documents is very low. So, analysis of factors becomes insufficient, and many prevention measures cannot defeat the latent factors. Therefore, we proposed an incident report support system. Next, we will want to improve the incident report support system to better support medical staffs.

REFFERENCES

Bello G.C. and Colombori V 1980. The Human Factors in Risk Analyses of Process Plants : The Control Room Operator Model (TESEO), Reliability Engineering,

FARMER F.R 1980. Reliability Engineering pp3-14

Hiromitsu Kumamoto, Ernest J. Hennley 1996. Probabilistic risk assessment and management for engineers and scientists: IEEE Press

Stuart K.Card, Thomas P.Moran, Allen Newel 1983. The psychology of human-computer interaction : Hillsdale, N.J.:L. Erlbaum Associates

Swain, A. D., Guttman, H. E 1983. Handbook of human reliability analysis with emphasis on nuclear power plant applications. : NUREG/CR-1278 (Washington D.C.)

Williams, J.C. 1985. HEART – A proposed method for achieving high reliability in process operation by means of human factors engineering technology in Proceedings of a Symposium on the Achievement of Reliability in Operating Plant, Safety and Reliability Society : NEC, Birmingham

Section V

Cognitive Engineering: Applications

CHAPTER 20

Promoting Temporal Awareness for Dynamic Decision Making in Command and Control

Sébastien Tremblay and François Vachon
École de psychologie, Université Laval, Québec, Canada
{Sebastien.Tremblay, Francois.Vachon} @psy.ulaval.ca

Robert Rousseau
C3 s.e.n.c., Québec, Canada
robert.rousseau@c3senc.ca

Richard Breton
Defence RD Canada – Valcartier, Québec, Canada
Richard.Breton@drdc-rddc.gc.ca

ABSTRACT

Temporal awareness is key to dynamic decision making (DDM) in a wide range of complex situations in safety-critical environments. Yet little is known about the effectiveness of temporal support when making critical time-based decisions. In the present study, we examine whether a decision support system (DSS) that provides an external time-based representation of the current situation and its evolution can promote temporal awareness in a C2 context. In a functional simulation of naval anti-air warfare, a baseline condition was compared to a condition in which a temporal overview display (TOD) taking the form of a scheduling aid was integrated to the original interface. The TOD was associated with an increased situation awareness level and higher concordance to a time-based decision heuristic relative to a distance-based heuristic, suggesting that the tool was successful in promoting temporal awareness. Still, performance impairment related to a higher propensity to miss critical changes to the situation was found in the TOD condition, possibly attributable to the need to process an additional display in the interface. Whereas these findings provide encouraging support for the benefit of explicit temporal representation in the effective processing of time-based information, they also call for more holistic evaluations of the cognitive impacts of DSS on DDM.

Keyword: Temporal awareness; dynamic decision making; command and control; scheduling aid; decision support system

1   INTRODUCTION

For a majority of tasks performed in the real world, in particular for the control of complex situations in safety-critical environments, real-time dynamic decision making (DDM) is required (Lerch & Harter, 2001). There is a consensus over the four basic characteristics of tasks that involve dynamic decision making (DDM): 1) Such tasks require a series of decisions and actions rather than a single decision; 2) decisions are not independent; 3) besides the impact of decisions, the situation changes on its own; and 4) decisions occur in real-time. In DDM, time is a central concern (e.g., Brehmer, 2005; Gonzalez, 2005). Ariely and Zakay (2001) point out that the dimension of time in DDM may refer to the time taken to make a decision, the tempo—i.e. the rate of changes in the environment—as well as the timeliness of the series of decisions. An important part of the complexity of DDM comes from the interaction between different aspects of the temporal dimensions. Such temporal interactions make DDM a very challenging endeavor that can greatly benefit from support. The objective of the present study is to test whether the use of a temporal overview display (TOD) in addition to geospatial displays would facilitate the exploitation of time in DDM, and could be beneficial on task performance.

Although time is a key dimension of DDM, there is little work on its contribution to DDM and even less on its support. In Endsley's (2000) model of situation awareness (SA) one can argue that the processing of events in time plays an essential role. Indeed, the third stage of SA involves a projection and extrapolation of the status of objects in the near future. While this aspect of Endsley's SA model recognizes implicitly the role of time in the projection process, Grosjean and Terrier (1999) have gone a step further and coined the term *temporal awareness* to emphasize the specific aspect of time in SA. Temporal awareness refers to the quality of the mental representation of the recent events and of what will or could occur next. In the present study we examine the role of temporal information to DDM in the context of naval command and control (C2). In the dynamic, complex and information rich environment of so-called maritime DDM and naval operations, the need to augment human performance at the individual and team levels is critical. One of the key cognitive requirements underlying maritime DDM is the process of temporal information.

Traditional geospatial displays have a potential drawback, even when designed properly: There is no explicit representation of temporal information (e.g., Bennett,

Payne, & Walters, 2005). The key advantage of a temporal display is that it goes beyond the mere snapshot of a situation in time (Endsley, 2000). For instance, some dynamic patterns may be extracted through examination of a temporal display that might go unnoticed in the geospace. At the same time, it is a challenge to portray temporal information outside of the geospatial context, with no reference to space. For example, Rousseau et al. (2007), in the context of simulated naval weapon-target scheduling, showed that a temporal display promoted the use of appropriate time-based heuristics for DDM but only after substantial practice. Hence, some researchers in visual analytics and developers recommend that geospatial and temporal data should be used in a complementary manner (e.g., Chalmers, 2003; Donnelly, Bolia & Wampler, 2007). Yet, it can be very difficult to consider all other situational dimensions when visualizing time-oriented data (Aigner et al., 2007).

DDM has been investigated with great success in laboratory settings through the use of microworlds, which represent a good compromise between experimental control and realism (e.g., Brehmer, 2005; Gonzalez, Valyukov, & Martin, 2005). The task studied here is a simplification of threat evaluation and weapon assignment in naval anti-air warfare. We used the S-CCS microworld (e.g., Vachon et al., in press) in which participants have to monitor a radar screen representing the airspace around the ship, evaluate the threat level and immediacy of every aircraft moving in the vicinity of the ship based on a list of parameters, and take appropriate defensive measures against hostile aircraft. The threat immediacy judgment and the application of combat power can be viewed as time-based decisions. Assessing threat immediacy—i.e. the temporal distance of a threat from the ship— relies on the evaluation of a key dimension in naval C2, namely the time-to-closest point-of-approach (TCPA; Roy, Paradis, & Allouche, 2002) which refers to the time the aircraft will take to reach the CPA if its velocity remains constant. In the same vein, weapon assignment consists in scheduling the sequential application of combat power based on threat immediacy. The aim of the present study is to determine whether such time-based DDM can benefit from a decision support system (DSS) that provides a time-scaled event-based representation of the situation.

A baseline condition is compared to a condition in which a TOD taking the form of a scheduling aid was integrated to the original interface to support temporal awareness. The TOD explicitly represents essential temporal components such as time-to-decide and time to hit a target by presenting each aircraft across a timeline. In order to assess the support provided by a TOD we measure DDM efficiency in terms of accuracy and speed of threat immediacy judgments as well as the quality of weapon-target assignment. In addition to metrics of defensive effectiveness, one key measure is the proportion of TCPA concordance indicating the degree to which target selection follows the rule that the aircraft with the lowest TCPA should be neutralized first. Adopting a holistic evaluation approach designed to identify benefits as well as potential cognitive costs to using the time-based system (Lafond et al., 2010; Vachon et al., 2011), we extract further metrics relating to the level of SA and other performance aspects. We also monitor eye movements as an index of attentional allocation to determine to what extent the TOD is exploited by operators.

2 METHOD

2.1 Participants

Forty-three students from Université Laval reporting normal or corrected-to-normal vision received a monetary compensation for their participation in a single two-hour experimental session. Twenty one of them were randomly assigned to the No-TOD condition and 22 to the TOD condition.

2.2 Microworld

The S-CCS microworld is a low-level, computer-controlled simulation of single ship naval anti-air warfare. The simulation is dynamic and evolves according to a scenario in interaction with the operator's actions. Typical scenarios of the S-CCS microworld involve multiple aircraft moving in the vicinity of the ship with possible attacks requiring retaliatory missile firing from the ship. A single participant plays the role of tactical coordinator who must be sensitive to changes in the operational space, conduct threat assessments including the categorization and prioritization of threats, and schedule the application of combat power.

Two variants of S-CCS were compared: a control system involving no specific DSS and a system with a TOD. In the No-TOD condition, S-CCS is presented in its simplest version. Its basic visual interface (see Figure 1) can be divided into 3 parts: 1) a tactical geospatial display, 2) an aircraft-parameter text list, and 3) a set of action buttons. The geospatial display is a type of radar display representing in real time all tracks surrounding the ship (represented by the central point). An aircraft can appear anywhere on the radar screen as an icon with a white dot surrounded by a green square. Aircraft move on the radar screen at various speeds and directions represented by a leader connected to the square. The direction of an aircraft can change according to a scenario. To select a particular aircraft, participants must click on its icon with the mouse. When selected, the square surrounding the aircraft turns red. The parameters list provides information on a number of parameters about the selected aircraft. The action buttons allow participants to determine threat level and immediacy of aircraft and to engage a candidate target.

In the TOD condition, a TOD was added to the original interface. This TOD, located to the right of the geospatial display (see Figure 1) represents temporal information explicitly by presenting each aircraft across a timeline. In fact, the TOD presents the same information about speed, distance, and direction as the geospatial display, but in a time-scaled event-based representation. However such a representation is more abstract in as much as it does not visually represent objects in their physical environment. The temporal interface is made of a grid background where vertical lines delimited specific temporal intervals. A single vertical red line identifies the "now". Each aircraft is represented by a horizontal rectangle, moving from right to left. An aircraft hits the ship when the right end of its rectangle crosses the red line. Every action made on the geospatial display occurs simultaneously in the TOD, and vice versa.

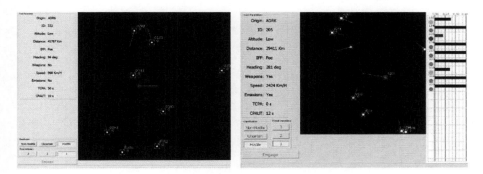

Figure 1. Screenshot of the S-CCS interface in the control (left) and TOD conditions (right).

2.2.1 Task

As tactical coordinators, participants performed three tasks. First, they assessed the level of threat posed by an aircraft by classifying all aircrafts as non-hostile, uncertain, or hostile based on whether critical cues have a threatening or non-threatening value. The number of threatening cues associated with an aircraft determines its threat level. When a threat level judgment was made, participants were required to click on the corresponding button. However, since aircraft threat level could change over time in a scenario, participants had to check the parameters regularly in order to re-assess the threat level, if appropriate. Status changes whereby an aircraft becomes hostile were considered as critical because hostile aircrafts were programmed to hit the ship. A total of 8 critical changes occurred unexpectedly in each scenario. Each critical change was accompanied by either an increase of speed, a change of direction (i.e., the aircraft is heading towards the ship), or both. A critical change was considered detected if an action was taken on the aircraft within the 15 s following this change.

The second task comprised assessing the threat immediacy of all aircraft designated as hostile based of their spatio-temporal proximity to the ship. In the No-TOD condition, threat immediacy had to be evaluated by adding the parameter values for TCPA and CPAUT (CPA distance in units of time, i.e. the time it will take the aircraft to hit the ship from the CPA; see Roy et al. 2002). In the TOD condition, such information was directly provided in the TOD. Participants had to determine whether the threat-immediacy level was high, medium, or low, and to click on the corresponding immediacy button.

The third task was to defend the ship against hostile aircraft in a timely manner by launching an anti-missile using the 'Engage' button when the target was within range. A small white dot, representing the anti-missile device, then appeared on the radar screen and homed at the target. Only a single target could be engaged at a time while more than one aircraft can attack the ship at once.

2.3 Procedure

Following a tutorial describing the context and the tasks to execute, participants were required to perform 2 training sessions, each composed of 4 4-min scenarios in order to familiarize them with the microworld's dynamic environment. There were 4 experimental sessions counterbalanced across participants. In each experimental session, 4 scenarios lasting 4 min each were performed. Each scenario involved a set of 27 aircraft varying in speed and trajectory on the radar screen. A maximum of 10 aircraft could appear on the radar screen at the same time.

In order to assess the level of SA during scenarios, we used the Quantitative Analysis of Situational Awareness (QUASA) technique (McGuiness, 2004), which combines both objective SA from accuracy of responses to queries (true/false probes) about the situation and subjective SA from self-ratings of confidence for each probe response. In every scenario, participants underwent a 24-s interruption period during which the whole S-CCS interface disappeared and was replaced by 3 consecutive probes. These probes covered all three levels of SA (Endsley, 2000). Note that the scenario continued to evolve in real time during the interruption.

3 RESULTS

3.1 System Usability

In order to determine whether participants used the TOD appropriately, we extracted eye-movement data recorded with a Tobii T1750 eye tracker at a sampling rate of 50 Hz. We computed 'on target' ratios (Poole, Ball, & Phillips, 2004), that is, the number of eye fixations on a particular area of interest (AOI) divided by the number of fixations over all AOIs. The threshold to detect an eye fixation was set at 100 ms and the fixation field corresponds to a circle with a 30-pixel radius. We focused the analysis to the threat-immediacy assessment period. In the No-TOD condition, the 2 most fixated AOIs during immediacy evaluation were the immediacy action buttons (29.6%) and the parameters relevant to the immediacy judgment (20.9%). With the TOD, fixations over the immediacy buttons slightly increased to 35.3% whereas those on relevant parameters dropped to 6.0%, most likely because 19.0% of fixations were made on the TOD. Such findings suggest that participants used the TOD appropriately during time-based decisions.

3.2 Decision Heuristics

The extent to which the use of TOD promoted TCPA-based decision making over the use of other decision heuristics (e.g., distance) was assessed using the proportion of concordance. Concordance refers to the extent to which the operator engaged missiles on the basis of the lowest TCPA (*TCPA concordance*) or the shortest distance (*distance concordance*). A concordance value of 1 indicates that each decision was optimal. Given that the 2 concordance measures were not fully independent (e.g., the aircraft with the lowest TCPA could also be the one with the shortest distance), we contrasted them using a ratio of their respective values. A ratio higher than 1 indicates that a TCPA heuristic was favored over a distance heuristic. The ratio of 1.06 found in the No-TOD condition significantly increased to 1.11 in the TOD condition, $t(41) = 2.67$, $p = .011$, suggesting that the DSS promoted the use of a time-based decision heuristic relative to a distance-based heuristic when scheduling engagement of hostile aircraft (cf. Rousseau et al., 2007).

3.3 Situation Awareness

We examined the impact of the TOD on SA level as measured by the QUASA technique. The results are shown in Table 1. Compared to the No-TOD condition, participants with the TOD were significantly more accurate on SA probes, more sensitive—showing both an increased hit rate and reduced false alarms— and better calibrated—the perception of their accuracy level was closer to their actual accuracy level so that were less overconfident. This pattern of results cannot be explained by a change in response bias. Such findings suggest that the TOD increased SA level.

Table 1 Summary of QUASA results (mean and SE) for the two conditions.

QUASA variable	No-TOD	TOD	$t(41)$	p
Hit rate	.606 (.029)	.697 (.026)	2.36	.023*
False alarm rate	.621 (.031)	.485 (.039)	-2.74	.009*
Sensitivity (d')	-0.54 (.124)	0.58 (.157)	3.16	.003*
Response bias (C)	-0.31 (.059)	-0.26 (.056)	0.61	.549
Accuracy level	.508 (.021)	.602 (.028)	2.69	.010*
Calibration bias	-.170 (.029)	-.065 (.034)	2.33	.025*

3.4 Behavioral Performance

In order to assess the impact of the addition of the TOD on performance, we extracted metrics for each of the main 3 tasks—determine threat level and immediacy, detect critical changes, and take defensive measures against hostile aircraft—and compared each metric across the 2 conditions using independent-samples t tests. All results are presented in Table 2.

We start with the 2 time-based processes the TOD was designed to support: judging threat immediacy and scheduling defensive measures. With regards to threat immediacy, although the TOD did not improve immediacy judgments in terms of accuracy, it significantly reduced the time required to make these judgments. These results suggest that providing a time-based representation of the current situation can speed time-based DDM.

However, the impact of the TOD on the application of the defensive measures was rather negative. Indeed, a first analysis revealed that the percentage of aircraft that failed to be neutralized before hitting the ship tended to increase in the presence of the TOD. We also extracted a defensive effectiveness metric related to how close the ship came to being hit. This measure is defined as the sum of the time-to-ship value for all hostile aircraft when either neutralized or when they hit the ship. A total of 0 means that all hostile aircraft hit the ship. The total is then divided by the number of hostile aircraft in order to obtain an average time-to-ship value. Higher values indicate a greater defensive effectiveness. The analysis showed that defensive effectiveness was significantly lower in the presence of the TOD, suggesting that the addition of the TOD seemed to hinder effectiveness in the scheduling of combat power, a process it was supposed to support.

Such drawback from the TOD on scheduling could be the consequence of the negative impact of the tool on another variable: the percentage of undetected critical changes. Indeed, significantly more critical changes were missed in the TOD than in the No-TOD condition. Given that this tendency to miss critical changes (i.e. the presence of new hostile aircraft in the airspace) was strongly correlated to the percentage of ship hits ($r = .84$, $p < .001$) and the defensive effectiveness index ($r = -.70$, $p < .001$), it is possible that the TOD impeded the application of defensive measures because it rendered participants less sensitive to critical changes in the operational space. With regards to threat evaluation, the TOD had no impact on both the accuracy and speed of the threat evaluation judgment.

Table 2 Summary of behavioral results (mean and SE) for the two conditions.

Variable	No-TOD	TOD	$t(41)$	p
Immediacy accuracy (%)	83.0 (2.5)	78.8 (4.3)	0.84	.408
Immediacy speed (ms)	1550.6 (73.7)	1327.4 (44.5)	2.62	.012*
Ship hits (%)	6.3 (0.7)	9.2 (1.3)	-1.93	.061
Defensive effectiveness (ms)	9200.6 (268)	7978.8 (293)	3.06	.004*
Undetected changes (%)	13.1 (1.1)	18.9 (1.9)	-2.61	.013*
Threat level accuracy (%)	86.0 (1.9)	85.1 (2.1)	0.33	.745
Threat level speed (ms)	2839.8 (130)	2952.1 (132)	-0.61	.549

4 DISCUSSION

We examined whether a DSS that provides an external time-based representation of the situation and its evolution can promote temporal awareness in the context of maritime DDM. Hollnagel (2001) emphasized the importance of the temporal dimension in complex and dynamic environments: "Control can be lost if the available resources—and especially time—are insufficient to evaluate the current situation and select the next actions" (p.246). Our results showed that providing a representation of key temporal information translated into faster threat assessment and improved SA, suggesting that the TOD was successful in promoting temporal awareness. The addition of a TOD seemed to foster the adoption of an effective TCPA-based selection rule. Moreover, the current findings suggest that there is value added for decision makers placed in dynamic situations to construct an optimal level of temporal awareness (Grosjean & Terrier, 1999).

Still, measures of defensive effectiveness against hostile aircrafts indicate that the TOD can nevertheless disrupt other aspects of the cognitive work. In fact, the scheduling of combat power application appeared hampered when the interface was augmented with the TOD (which was precisely designed to support this very function). Given the beneficial impacts of the TOD on threat-immediacy evaluation and on SA, it would be surprising that it could at the same time disrupt another time-based process. The observed impairment of combat power scheduling is likely to be related to the greater propensity to miss critical changes in the TOD condition. That inability to detect critical changes - change blindness – is related to poor performance in a threat assessment task (see Vachon et al., in press). This increased tendency to miss critical changes in the TOD condition is possibly attributable to the need to process an additional display in interface that would momentarily attract the attentional focus away from the geospatial display and the parameter list. Overall, the present study reveals a tradeoff between the benefits and overhead of supporting temporal awareness that calls for more holistic evaluations of the impacts of support technologies on DDM (Lafond et al., 2010; Vachon et al., 2011).

Our findings also provide insights into the development of support for the processing of temporal information. Over the last decades or so, there has been a great deal of work in the development of visualization techniques. Various methods have been proposed for the visualization of time-oriented data and for facilitating the analysis of such data. A range of techniques has been developed in order to capture the different aspects of time in a situation: for instance, whether the temporal dimension of the represented situation is about the linear or cyclic evolution of events or is concerned with discrete time points or time intervals. In the case of scheduling actions within a fast-paced dynamic environment, the most common visualization technique – and the one employed in the present study – is the use of planning bars that can provide information about when it is best to act (see Aigner et al., 2007). Our user-centered method of investigation combined with cognitive metrics has proved very useful in providing an objective, behavioral assessment of the costs and benefits of different visualization techniques.

ACKNOWLEDGEMENTS

We are thankful to Julie Champagne, Sergei Smolov, Thierry Moisan, and Benoît Vallières for assistance in programming, data collection, and analysis. This work was supported by a R&D grant from the National Sciences and Engineering Research Council of Canada with Defence R&D Canada, and Thales Canada.

REFERENCES

Aigner, W., S. Miksch, W. Müller, H. Schumann, and C. Tominski. 2007. Visualizing time-oriented data—A systematic view. *Computers & Graphics* 31: 401-409.
Ariely, D., and D. Zakay. 2001. A timely account of the role of duration in decision making. *Acta Psychologica* 108: 187-207.
Bennett, K.B., M. Payne, and B. Walters. 2005. An evaluation of a "Time Tunnel" display format for the presentation of temporal information. *Human Factors* 47: 342-359.
Brehmer, B. 2005. Micro-worlds and the circular relation between people and their environment. *Theoretical Issues in Ergonomics Science* 6: 73-94.
Chalmers, B. A. 2003. Supporting threat response management in a tactical naval environment. *Proceedings of the 8th ICCRTS*. Washington, DC.
Donnelly, B.P., R.S. Bolia, & J.L. Wampler. 2007. Capturing commander's intent in user interfaces for network-centric operations. *Proceedings of the 12th ICCRTS*. Newport, RI.
Endsley, M. R. 2000. Theoretical underpinnings of Situation Awareness: A critical review. In *Situation Awareness Analysis and Measurement*, eds. M. R. Endsley and D.J. Garland. pp. 3-28. Mahwah, NJ: Lawrence Erlbaum Associates.
Gonzalez, C. 2005. Decision support for real-time, dynamic decision-making tasks. *Organizational Behavior and Human Decision Process* 96: 142-154.
Gonzalez, C., P. Vanyukov, and M. K. Martin. 2005. The use of microworlds to study dynamic decision making. *Computers in Human Behavior* 21: 273-286.
Grosjean, V. and P. Terrier. 1999. Temporal awareness: Pivotal in performance? *Ergonomics* 42: 1443-1456.
Hollnagel, E. 2001. Time and control in joint human-machine systems. *Proceedings of the 2nd IEE People in Control Conference* CP481: 246-253.
Lafond, D., F. Vachon, R. Rousseau, and T. Tremblay. 2010. A cognitive and holistic approach to developing metrics for decision support in C2. In *Advances in Cognitive Ergonomics*, eds. D. B. Kaber and G. Boy, pp. 65-73. CRC Press.
Lerch, F. J., and D. E. Harter. 2001. Cognitive support for real-time dynamic decision-making. *Information Systems research* 12: 63-82.
McGuinness, B. 2004. *Quantitative Analysis of Situational Awareness (QUASA): Applying Signal Detection Theory to True/False Probes and Self-Ratings*. 2004 command and control research and technology symposium, San Diego, CA.
Poole, A., L. J. Ball, and P. Phillips. 2004. In search of salience: A response time and eye movement analysis of bookmark recognition. In People and Computers XVIII-Design for Life: Proceedings of HCI 2004, eds. S. Fincher, P. Markopolous, D. Moore, and R. Ruddle. London: Springer-Verlag.
Rousseau, R., S. Tremblay, D. Lafond, F. Vachon, and R. Breton. 2007. Assessing temporal support for dynamic decision making in C2. *Proceedings of the Human Factors and Ergonomics Society 51st Annual Meeting*: 1259-1262. Santa Monica, CA: HFES.
Roy, J., S. Paradis, and M. Allouche. 2002. Threat evaluation for impact assessment in situation analysis systems. In *Signal Processing, Sensor Fusion, and Target Recognition XI,* ed. I. Kadar. Proceedings of SPIE, 4729: 329-341. Orlando, FL: SPIE Press.
Vachon, F., D. Lafond, B. R. Vallières, R. Rousseau, and S. Tremblay. 2011. Supporting situation awareness: A tradeoff between benefits and overhead. *Proceedings of the 1st CogSIMA Conference*: 282-289. Miami Beach, FL: IEEE.
Vachon, F., B. R. Vallières, D. M. Jones, and S. Tremblay. In press. Non-explicit change detection in complex dynamic settings: What eye movements reveal. *Human Factors*.

CHAPTER 21

Data-driven Analysis and Modeling of Emergency Medical Service Process

Liang Wang, Barrette Caldwell
IE department, Purdue University
bscaldwell@purdue.edu

ABSTRACT

Emergency Medical Service (EMS) is an integrated part of the modern healthcare system. The major participants include schedulers and first responders, who communicate intermittently during the whole service procedure, from the "First response" stage to "Last clear". With data obtained from a Computer Aided Dispatching system of a university based EMS provider, data analysis results revealed issues in resource allocation and possible contributing factors. Uncertainty in the dynamics of event progression and situation awareness influences the performance of all participants in the EMS system. With respect to this dynamic feature of first responders' working scenario, the increased cognitive load of EMS responders, as well as their working behaviors, should be accordingly taken into account in the dispatching decision-making process. A preliminary conceptual model based on empirical data led to development of a new behavior model of first responders, summarized as an Information Concentration model. This theoretical model serves as an indicator of system performance during the response process after initial dispatch. Parameters included in this model can be derived from archival data of service performance from dispatch records and potential traffic data from GPS systems or other available traffic flow models.

Key words: EMS, computer-aided dispatch, information flow, event response, first responders, cognitive workload, information concentration

1 INTRODUCTION

A critical element of healthcare delivery, an important entry pathway to care, is the Emergency Medical Service (EMS) first responder system. EMS is a time-based rather than location-based system that spans from event start with emergency 911 calls to the discharge of treated patients at the accident scene or transfer to hospital emergency delivery. Generally speaking, the duration of the whole EMS process could range from several minutes to a few hours. As a result, it is difficult to predict final system performance based on all information available at the beginning stage.

Time pressure is an unavoidable concern in EMS response, as patients' survival and health outcomes are closely linked to response time (Blackwell 2002, pp. 288-295). In addition, a dispatching order demands a great deal of human effort, physical and cognitive, to make it work. First responders have to work under great time pressure and limited information; the original dispatch order cannot not assumed as correct. What is worse, as a large complex system with uncertainty and an expectation to operate in a variety of degraded situations, the EMS's dispatching system could even be damaged, resulting in catastrophic failures. The most notorious one, the London ambulance service failure in 1992, led to a reported 11 hour delay in dispatching and resulted in costs of 2.5M dollars and 20 lives (Finkelstein and Dowell 1996). The success and failure of EMS doesn't only affect the wellbeing of the people who are in need of help, but also those who try to help. Every year, the rate of ambulance crashes and EMS worker injuries is significant. In fact, the death rate of EMS personnel in the U.S. while at work is estimated to be as high as 12.7 per 100,000 EMS workers (CDC 1996, p. 52).

It is important to notice that these situations result from the challenges of managing EMS as a dynamic, safety-critical system. To minimize response time without risking first responders' lives, more factors rather than the general length of path of travel, should be taken into account. Based on task analysis and data-based analyses, more variables could contribute to the dynamic features of EMS systems and design of first responder interfaces.

Based on these considerations, a study was conducted in a University-affiliated EMS system to find out how and how much elements of information flow and the dynamic situation influence the performance of this time-critical system. Data for this study included interviews of the Chief of this EMS agency, the related hospital and first-responders, and quantitative log data from the EMS Computer Aided Dispatching system.

2 INFORMATION FLOW VARIABLES IN EMS SYSTEM

An EMS system provides two types of services, differentiated by the nature of the emergency. For cases like transportation of patient from one facility to another, the ambulances are allowed to travel at a regular speed. For "alarm and siren" situations, the ambulances are required to travel at a high speed. The priority determination of services can be more or less helpful in dispatching strategy selection and the protection of first responders' safety (CDC 1996, p52), but this information is not always known in advance.

Based on Caldwell's terminology (Caldwell 2008, p427-438), EMS represents a distributed human supervisory coordination environment incorporating the dispatching center and the first responders, with possible inclusion of hospital emergency departments. First responders are expected to integrate three possible sources of information: the calling center, the accident scene and the dynamic traffic situation. For different types of services, the information from different sources should change from case to case, because they have different features with respect to time (Boustany 2011).

Not all independent variables are influential on the performance of the system. Those that contribute to generate and sustain appropriate situation awareness could be considered relevant.

According to Ensley's definition of situation awareness (SA), EMS first responders, especially ambulance drivers and emergency management technicians (EMTs) must maintain perception of the physical environment, dispatching information, and patient status; comprehension of current task and route demands, and projection of available routes, care capabilities, and handoff requirements as the ambulance transports the patient (Endsley and Garland 2000).

At the perception level, the signal detection phase, EMS participants are supposed to be aware of the surrounding situation elements. The ambulance staff on board should be aware of the traffic conditions and signals, including the surrounding vehicles, as they are driving on the public access roads with all other vehicles. At the comprehension level, the sensory-cognitive step, a series of decisions should be made to align their navigation and patient preparation with the overall task goal to stabilize and send the patient(s) to the hospital quickly and safely as possible.

EMS first responder sense-making activities are based on the environmental elements acquired at the perception and comprehension levels, as well as responder abilities, including their level of education, the skill set and the ability to process dynamic information under time pressure (Endsley and Garland 2000).

3 STUDY DESIGN AND RESULTS

Both the interview and archival data analysis for the current study were conducted focusing on the EMS functions of a university-based first responder system. This system incorporates campus and county operations, and has recently (as of mid 2009) implemented a Computer-Aided Dispatch system. These sources of data were used in support of a master's thesis project to examine cognitive and navigation demands of EMS first responders.

3.1 Campus EMS Context

There are two types of EMS workers as campus first responders, the Emergency Support Technician (EST) and the paramedic. Paramedics have more advanced training and are capable of providing a wider range of medical services. The campus has 2 fire trucks, 1 airport fire truck and 2 vehicles with Advanced Life Support (ALS). Normally, for each dispatched vehicle, there should be 1 paramedic and 1 EST on board, according to State law. However, the campus EMS vehicle usually has two paramedics on board. All first responder calls through the campus 911 system are categorized into unique types and denoted with a standardized National Fire Incident Reporting System (NFIRS) 3-digit number.

No route navigation information is provided by the dispatch center; only the location of the accident / event site is provided, which is illustrated as a dot in the computer-aided dispatch system interface. Other information, such as the detailed address, is also shown on the screen of the EMS vehicle in text format.

All drivers are obliged to remember all traffic situations and road conditions on campus, as well as changing demands associated with campus activity. They also experience road construction undertaken during the summer. EMS ambulance crews are not granted the right to drive with highest priority. Thus, they must obey stop signs and red lights, and must "ask for priority "on public roads or unusual routes, such as the sidewalk or other pedestrian areas.

The process of interviewing and data collection extended from June 4th to June 28th, 2011, with final acquisition of all 180 pages of dispatch data logs (covering all of calendar 2010) on July 7th, 2011. Statistical analysis was conducted to analyze the dispatch data, to determine a) the variation and range of information and physical resource dynamics for EMS responders to describe and facilitate the current computer aided dispatch system. b) Which features should be taken into account in constructing a dynamic computer simulation of the information flow, navigation, and workload demands of an evolving EMS event.

3.2 Data analysis and results

A primary aspect of the archival data analysis is to determine the time course and event progressions of EMS response dynamics, beginning with the dispatch call. According to Figure1 (see below), the nature of an emergency call varies greatly. In most situations, the information regarding the nature of the accident is not revealed until right after the dispatching order. So an independent variable, which could describe the possible outcome in accident scene, is required as a situation element.

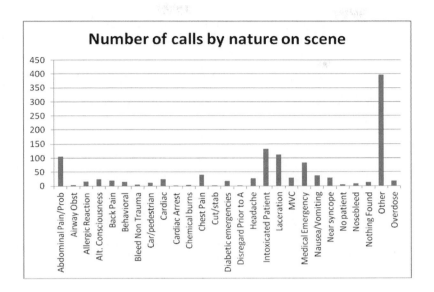

Figure 1: Distribution of EMS dispatch calls

In this collection of 2010 events, NFIRS Type 311 and Type 321 are the most frequent types of dispatch calls (as identified at the time of the call). These two event types represent emergencies at different levels of seriousness and importance (NFIRS, 2009, p. V).

In most real cases, EMS responders cannot determine the actual type of a certain incident until they arrive at the scene. Thus, the number of cases classified by the dispatch center will not necessarily map the incident report from the EMS responders (see Table 1). For instance, almost a quarter of all reported motor vehicle accidents turned out to be with no injuries and less urgent. These could be regarded as "false alarms," depending on the interpretation of the incident call. Although additional analysis and determination of these cases in terms of signal detection theory might be possible, the clear indication is that one cannot assume that the event projection at the dispatch call will necessarily be the event that is presented to the EMS first responders at the scene when they arrive. In the particular case of vehicle accidents and trauma, there are strong concerns that the cost of a slow response to a real accident (a signal "miss") has much broader and higher social costs than a rapid response to a less critical event (a signal "false alarm").

Table 1: Incident Case Types and Classifications at Dispatch and Response

NFIRS Incident Type		#Cases	
Code	Incident Statement	Dispatch	Response
311	Medical assist, assist EMS crew	187	182
321	EMS call, excluding vehicle accident with injury	405	391
322	Motor vehicle accident with injuries	14	0
323	Motor vehicle /Pedestrian accident	9	0
324	Motor vehicle accident with no injuries	4	0
	Total number of calls	615	573

EMS responders would prefer to assume that all reported motor vehicle accidents are very urgent, and then they will chose to rush to the accident scene in "siren and alarm" state. But according to the data analysis, almost 25% of times, they will face a "false alarm", which represents heightened risk and use of EMS resources.

A second question identified by both interview discussions and preliminary data analysis was whether ambulance response differed based on the type / level of NFIRS emergency? A two-sample t-test was conducted to compare the response times for Type 311 / 321 (see Table 2). The results of the t-test were not significant ($p=0.0952$), indicating that the mean response times of these two types of services are equal.

Table 2: Response Times (in sec) for Type 311 and 321 runs

	Type 311 (n=187)	Type 321 (n=405)
Mean	267.14	247.87
Variance	111314.3	163654.8

These results show that current EMS responders used similar levels of effort, including the travel speed and other preparations, for two types of calls at different emergency levels.

4   DISCUSSION AND CONCLUSION

Based on our interview and archival data analysis, we were able to develop a better understanding of EMS responder behavior and the dynamics of EMS events after the initial dispatch call. As an overall fraction of emergency event progression, dispatching and en route time is often small compared with the on scene and patient transport time. However, initial response time is a major determinant of the perceived quality of the EMS system.

Although the improvement in logistic planning could provide better selection of route at the beginning to potentially shorten the response time, the quality of EMS response is also dependent on the information available to the first responders throughout the event response. They require a reasonable understanding of the destination prior to arrival in order to make proper preparations. As currently structured, EMS system information does not necessarily provide this information to the response vehicle. The 2010 archival data reflects that there are no significant difference in response time based on event severity, and a limited predictability of actual response severity.

If more detailed and complete information about the accident scene were provided to responders, resources (including all gear and personnel), would be better allocated and safety of EMS responders could be better allocated and planned.

The last but not the least important conclusion is that the demands of different types of campus EMS fluctuate although the year. Developing more sensitive predictive models of information, physical environment, and task situation concentrations of workload demands could help campus and other EMS responders. The goals of this project are to examine quantitative measures of information concentration that more effectively support EMS first responders to improve their working performance and lessen their cognitive workload during emergency event response.

REFERENCES

Blackwell, TH and Kaufman, JS 2002, 'Response Time Effectiveness: Comparison of Response Time and Survival in an Urban Emergency Medical Services System' ,*Academic Emergency Medicine*, 9: 288–295. doi: 10.1197/aemj.9.4.288

Boustany, KC 2011, 'Impact Of Contextual Metadata On The Perceived Effectiveness And Efficiency Of Team Coordination Processes In Healthcare Operations,' West Lafayette, IN.

Caldwell, BS 2008, 'Knowledge sharing and expertise coordination of event response in organizations', *Applied Ergonomics*, 39,427-438.

CDC, 2003,'Ambulance Crash-Related Injuries Among Emergency Medical Services Workers - United States, 1991-2002', *MMWR* , 52(08), 154-156.// 14

Endsley, MR and Garland, DJ 2000, 'Theoretical Underpinnings of Situation Awareness: A Critical Review', In *Situation Awareness Analysis and Measurement*, Mahwah, NJ, Lawrence Erlbaum Associates.

Finkelstein, AD and Dowell, J 1996, 'A comedy of errors: the London Ambulance Service case study', *In 8th International Workshop on Software Specification and Design*

United States Fire Administration 2009, *National Fire Incident Reporting System (NFIRS 5.0) NFIRS Data Entry/Validation Tool Users Guide NFIRS 5.0 Software Version 5.6*, Department of Homeland Security ,Federal Emergency Management Agency, United States Fire Administration,

CHAPTER 22

Age and Comprehension of Written Medical Information on Drug Labels

Melvis N. N. Chafac,[1,2] Alan H.S. Chan[2]
[1]School of Industrial Engineering
Purdue University
West Lafayette
IN, U.S.A
mchafacn@purdue.edu
[2]Department of Systems Engineering and Engineering Management
City University of Hong Kong
Hong Kong

ABSTRACT

Recently there has been a surge of interest in medications and medication labels. Like pharmacists, most patients have to deal with medication (drug) labels on a daily basis. Health care systems worldwide are increasingly recognizing the need for a patient centered labeling system. But then, why are drug labels important to patients? Gerontological projections suggest a rapid increase in the number of old people such that a significant part of the patient population would be made of older adults. This research would explore the different studies conducted on the comprehension of drug labels. It would also analyze the problems, difficulties and errors of the elderly group in consumption of medicine, and if there is a need to change current drug labels for older patients.

Key words: older adults, drug label, comprehension

1   INTRODUCTION

In the last few decades, the U.S. has placed great importance on communication in health care, as a key to increasing comprehension of medical information. As early as 1968, Korsch, Gozzi and Francis noted the importance of doctor-patient communication. In 1995, Ong et al. termed communication "the main ingredient in medical care." Communication and comprehension are intricately linked, and often occur together. Besides the obvious importance of communication of medical information between medical professionals, there is also the need for effective communication of medical information to the consumers or patients. In order for effective communication to take place, the patients have to comprehend the information presented. Simply put, comprehension is associated with correct interpretation and correct execution of the information at hand. Communication in this setting is considered effective only when it produces the desired effect in the patient or audience. Age is one of the most important aspects that affect comprehension of medical information.

2 WRITTEN INFORMATION ON DRUG LABELS

2.1 Importance of written medical information on drug labels to consumers

One of the most consistent ways of communication of medical information to patients is through written material. Medical information is presented in many written documents such as the drug container label, consumer medicine information, the medication guide and the package insert. This paper focuses on the drug label as the single, most important source of written information for patients. The drug or medication label would include all the information presented on the medication containers, including all affixed stickers. The drug label is a reference for the patients every time they take the medication. In some cases, this has become the main or only source of information for the patients (Bailey et al., 2009). The drug label is very important because it contains pertinent information about the medication. It usually has a set of instructions that provide guidance on how the medication should be administered to the patients. The label specifies the correct dosage, times and methods of administration. That is, it tells the audience how many units to administer, when they should be administered and method of administration, as well as any necessary accompaniments such as food, water or milk etc.

Aside from this, the container label also plays a role in medication management. It identifies the medication, the patient for whom the prescription was written, the prescriber and the pharmacy (Vredenburgh and Zackowitz, 2009). Some labels include the manufacturing information such as the name of the manufacturer, batch number and the expiry date. Medication labels also specify the major components in the medication. This provides information necessary to understand the medication interactions (what other chemicals should not be taken in conjunction with the medication) and possible allergies. In most cases, warnings about the medication are presented on the container labels.

2.2 Factors that could lead to miscomprehension of drug medication label

Readability of the information presented on the medication label - Presentation of the information on the medication label is important. Using signals or attention grabbers (font, writing style, organization, color, etc), iconic information presentation (active vs. passive voice, use of poignant common terminology), could draw patient attention to important aspects on the label. Furthermore, ease of reading drug labels (readability) is impacted by the level of difficulty of the information presented (reading grade level). Parker (2000) noted that most written medical information is presented at a grade 10 reading level, while the average American adult reads at or below a 9th grade level. Thus, information on medication labels would be overwhelming, instead of helpful, to patients especially those with low health literacy.

Reading ability of the audience - In 1980, Eaton and Holloway found a significant connection between comprehension and reading ability. Sadly, there are about 44 million or more Americans with below the national standards of literacy (Quirk, 2000). That is, their inability to read and write well hinders their daily functionality and work lives. Given that normal literacy does not directly translate into health literacy, even more people in the U.S. have low health literacy. Low literacy is the "inability to read, write and use numbers effectively" (Pignone et al., 2005). Health literacy is thus the ability to read, write and use or interpret numbers associated with medical information effectively. In the National Assessment of Adult Literacy study conducted in 2003, it was noted that 36% of the 19000 participants had basic (could carry out simple everyday activities) or lower than basic health literacy (Kutner et al., 2007). In the U.S. it is estimated that about 90 million adults have limited literacy skills (Parker, 2000). This means that roughly more than half of the American adult population would have difficulty comprehending medication information. It is reasonable to infer that these people have difficulties accessing medical information, which may be pertinent to their health and to the prevention of negative outcomes.

3 AGE

A third of the health care expenditure in the U.S is spent on people aged 65 and above, and within the next 30 years, a 25% increase in health expenditure is expected (AHRQ and CDC, 2002). Due to significant medical advancements, some conditions that used to kill people are no longer as deadly and as such, many people live to a ripe old age. The elderly population in the U.S. is projected to increase from 39.6 million (as noted in 2009) to 72.1 million by 2030 (AoA, 2011). That is, about 19% of the population would be older adults. Living longer does not mean living ailment free. Roughly 80% of older adults in the U.S. live with one or more chronic illnesses (CDC and NCCDPHP, 2011).

3.2 Problems associated with age that affect comprehension of medical information

Older adults are the main consumers of medications in the U.S. (Coppard, Coover and Faulkner, 2011). According to the FDA report on medication errors, nearly half of the fatal errors recorded occurred in older adults (Meadows, 2003).

Physiological Changes. Many researchers have acknowledged the vulnerability of older adults due to a variety of aging changes (Coppard, Coover and Faulkner, 2011; Hastings, Kosmoski, and Moss, 2010; Coleman, Wagner, Grothaus et al., 1998). One inescapable aspect of human aging is the development of limitations such as reduced dexterity and speed, impaired vision, declined hearing ability and reduced cognitive ability (Taplay and Flores-Vela, 2012). With limited dexterity and movement, comes increased difficulty in opening medication containers, tearing medication bags or blister packs and halving tablets. Aside from this, vision impairment increases the difficulty in reading container labels. This is problematic because correct interpretation of visual cues such as color, shape and written information allows patients to correctly recognize the medication and the correct method of administration. About 17% of American adults aged 65 and above are visually impaired (Leonard, 2002; Caban, Lee, Gómez-Marín et al., 2005). Another common limitation faced by this group of patients (28% of American older adults) is declined hearing ability (Dey, 1997). Reduced hearing increases difficulties in understanding oral communication. This may reduce the patients' understanding of oral communication and make them more dependent on written communication such as drug labels.

Most cognitive impairments affect the patients' ability to comprehend medical information. The ability to recall is one of the most prominent functions that are affected by cognitive impairment in older adults. Normally, retention of medical information is quite short; prior research showed that patients immediately forgot 40-80 % of the information provided by the medical professional (Kressels, 2003). As people age, their ability to recall information declines (Ghetti and Bauer, 2011). Added to poor memory, age is inversely proportional to the amount of information that is correctly recalled. Thus, the older the patients, the less information they correctly recall (McGuire, 1996; Anderson, Dodman, Kopelman et al., 1979). This could lead to over-adherence whereby the patients forget that the medication was taken before, and takes it again; or unintentional under-adherence. An example of unintentional under-adherence is when the patient believes that they have taken the medications as prescribed for the day, when in fact, the last time the medications were administered was a day before.

Reading Ability- Besides the visual impairments that interfere with their visual acuity, older adults have limited reading, understanding and computational skills than younger adults (Rudd, Moeykens, and Colton, 1999). In a study by Gazmararian and her group (1999), it was noted that the reading ability of older adults decreased strikingly with age across all educational levels. This is not surprising, as cognitive impairment affects executive functions such as problem definition, idea generation and development, planning and performance of goal oriented activities (Hastings, Kosmoski, and Moss, 2010).

Polypharmacy - Most older adults suffer from more than one chronic illness and as such have to take multiple medications (polypharmacy). Some researchers such as Coppard, Coover and Faulkner (2011) have stressed the need for health care professionals to ensure that the patients can manage their medications correctly. Taking many medications can be confusing as there are many different doses to take at different times and with different means of administration. Additionally, older adults also have to make decisions about storage and refills and whether to tell the physician about new symptoms. Medication interaction is also an ever present danger of polypharmacy.

Misunderstanding of the treatment regimen - Older adults might not fully understand the need for some of the medications that they are given and so may tend to neglect them if no real effects can be noticed. A lack of understanding of the purpose of their medication and the associated risks permits some patients to self-prescribe. This sometimes leads to patients intentionally taking more than the prescribed dose in order to increase effectiveness or resorting to over-the-counter medications, which react adversely with the prescribed medications. Non-adherence could result from patient misunderstanding. Adherence is the degree to which the patient's actions conform to the guidance of the prescriber (Horne, 2006). 11% of hospitalizations in older adults is due to non-adherence and patients with chronic disease tend to have higher incidence of non-adherence to their regimen (Marek and Antle, 2008). Non-adherence could lead to relapse, medication interactions and could ultimately result in adverse drug events.

4   SEARCH METHODOLOGY

A search of words synonymous with 'health literacy,' 'medical information,' 'container label' and 'old age' in Medline, Cochrane and Google Scholar returned a vast amount of literature. This search suggested a surge of interest in health communication and aging. A more structured search in Google Scholar was conducted for articles related to 'Older adult comprehension of drug or medication labels.' This initially returned 6500 articles. This search returned 504 related articles (excluding patents) that had been published since 2011 to present (February, 2012). Extending the search to 2007 returned 1350 related articles. As shown in Figure 1, after screening out unrelated titles, there were 179 potentially relevant articles. Abstracts were then read and more articles were screened out. Disqualified articles were those that were irrelevant, were not based on a U.S population, were not in English, or were applicable only to a very specific population such as older adults with prostate cancer. This left us with 84 potentially relevant articles. These articles were then classified based on the main topics discussed. Of these, 28 were about literacy or health literacy, 25 were about drug medication labels or warning stickers on medication containers, 13 were about older adults and their medications, 10 were on aging, 7 were about literacy and aging or literacy and older adults' medical practices and one article was on literacy and medication container labels. A preliminary reading of the articles led to 13 articles with relevant content and data. From these 13, the authors chose the 3 most relevant based on content and application of rigorous research methods.

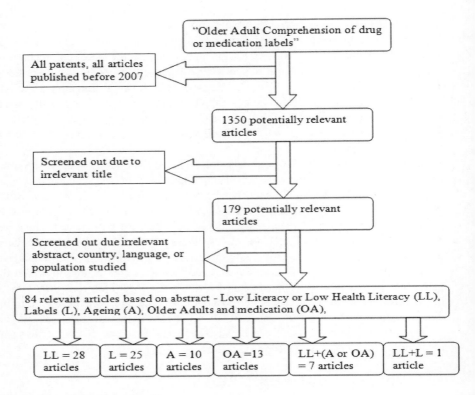

Figure 1 Search criteria and results

In their article in 2007, Shrank and Avorn singled out elderly patients (60% or more of those in the study) as having the most difficulty reading and understanding drug labels. They noted that medication information presented on labels is an important source of knowledge to patients. Patient education about prescribed medication is inefficient, compelling patients to rely on medication labels (and or medication inserts). Most of the patients who get information from health professionals about their medications forget and have to use their medication labels or other written information as a reference. Improving written information is worthwhile in order to promote safe and appropriate use of medications. These authors identified health literacy and medication errors as factors that affect understanding of written information. Their article describes several studies in which almost half of the patients could not correctly interpret drug label instructions and warning stickers.

Of the 1.5 million medication errors noted yearly in the U.S., inefficient labeling was noted as a critical source of medication error. Poor understanding of medication instructions inadvertently provides a barrier to compliance. Shrank and Avorn observed that though the Food and Drug Administration (FDA) and state boards of pharmacy jointly regulate the content of container labels, few regulations have been established regulating the format of medication labels. Existing data suggests that medication information is presented in ways that are increasingly complex and of low quality. This poses a problem for most Americans but even more so for older adults. The authors noted a paucity in data about medication labels, though casual observation suggests great variability in content and format of labels. The authors concluded by expressing a need for easily legible, clear and distinct labels and warning stickers (Shrank and Avorn, 2007).

A 2008 article by Marek and Antle noted that older adults spent more than 14 billion per year on medication related expenses. They found that older adults in community dwellings faced problems such as medication adherence and complex medication regimens. Non-adherence increased with increasing age group and increasing clinical complexity of care. The authors suggested that one of the main things that could be done to aid elderly adults was to reconcile their medications to ensure that they are taking what the physician ordered and no more or no less than was ordered. This would reduce the number of negative medication interactions. Other interventions that have been found to increase compliance in older adults include the development of programs for pharmacy delivery, reminders to take medications and to get refills, as well programs to manage the financial burden of the medications such as health insurance.

Marek and Antle reported studies in which less than 25% of the older adults knew the consequences of drug omission or the toxic side effects. Additionally, most older adults had insufficient knowledge of how to administer their medications. Patient education is thus necessary. Inability to read medication labels has been linked to non-adherence in elderly patients. 47% of the patients in a community study reported having difficulty with the medication labels due to poor vision, language reading barrier, and small writing. Older adults have been more responsive to logically ordered, explicit and listed items. Cognitive ability of older adults was noted to affect medication management. The authors suggested that setting aside a fixed time, location and routine would be beneficial for older adults. Intentional non-adherence was also observed in older adult patients, and the need for continuous monitoring to ensure patient compliance and avoid adverse drug events was noted. The authors concluded by suggesting guidelines for medication management for older adults (Marek and Antle, 2008).

Bailey and her group in 2009 noted that inefficient medication labeling is likely a significant cause of medication error. Labels are the most palpable and easily obtainable source of medication information. The authors pointed out that existing drug labels are not user-friendly, are vague and ambiguous and usually only in English. Improving medication labels could potentially lead to increased safe and appropriate medication use. The drug label is often the only source of written medical instructions for the patients. Improving labels are thus an important initial step toward patient understanding of medication instructions given that other sources are inadequate. Physicians and pharmacists often fail to communicate detailed information about the prescribed regimens in ways that support patient compliance, and other written sources of information are written at high reading grade levels and are not specifically designed for patients. The Internet, though a valuable resource, provides information which is hard to validate and often incomplete.

Though important, significant issues detract from the efficiency of medication labels. Labeling standards vary from pharmacy to pharmacy, vary based on what the prescriber writes and vary based on how these written directions are interpreted by the pharmacist. Format of presentation of the information on medication information also adds to the complexity. As noted before, low health literacy affects comprehension of medication information and is often the plight of older adults and minorities. Bailey and her group also postulate that language barriers are a significant problem, as 44 million adults in the U.S. have limited English proficiency (LEP). The authors provide ample evidence of the strong link between LEP and misunderstanding of medication labels. They suggest that providing medication instructions on the label in the language of the patient would reduce medication errors. Bailey and her partners suggest an organized presentation of simplified text and use of numeric characters as opposed to words to specify the dosage. Larger texts and typographic cues, clear and concise content and making considerations for patients with LEP are other methods to enhance patient comprehension (Bailey, Shrank, Parker et al., 2009).

To the best of our knowledge, there are hardly any recent articles that study the comprehension of medication container labels by older adults. An earlier study (in 1985) of 60 older adults reported that 60% could not read the medication labels and about 60% were unclear about the instructions on the labels (Zuccollo and Liddell, 1985). Another research fairly relevant to the topic is that by Smither and Braun, published in 1994. They studied the legibility of drug labels to older and younger adults and found that older adults had more difficulties reading the labels than younger adults (Smither and Braun, 1994).

4.2 Consequences of miscomprehension of the drug label

According to Shrank and his group, "medication errors are an important public health concern and poor understanding of medication labels are a root cause" (Shrank, Parker, Davis et al., 2010). Based on the report by Stoppler and Marks (2009), roughly 1.3 million of the injuries encountered annually in the U.S. can be attributed to medication errors. A medication error is any event that could cause or is linked to inappropriate medication use or patient harm and is preventable. These events could be related to several aspects of health care including but not limited to drug container labels, medication monitoring and use (FDA, 2011a). It is hard to completely eradicate the occurrence of medication errors in health care but it is possible to reduce the number of these events, as some of them are due to misinterpretation of instructions on the drug label, and inability to correctly distinguish similar labels (Chermak, 2009). Based on an earlier study by the FDA, improper dose administration (44%), wrong drug (16%) and wrong route of administration (16%) were among the leading fatal medication errors (FDA, 2011b). In a study of older adults, it was noted that 10.6% of Emergency Department (ED) visits were related to adverse drug events and 31% had one or more potential adverse drug interactions based on their medications (Marek and Antle, 2008).

Low quality of life can be associated with miscomprehension of medication labels. Patients with low health literacy have lower quality of life as they are unable to read and comprehend important health information, which could be helpful in avoiding mishaps. Low health literacy limits access to health information as the patients find it difficult to comprehend written and sometimes oral information about preventive methods, warnings and drug interactions.

5 CONCLUSION

Drug labels play a critical role in communication of medical information to patients. As noted above, several inefficiencies in medication container labels, as well as limitations in the patient's ability detract from the comprehension and safe use of medications by patients. A 1993 to 1998 study conducted by the FDA found that up to 44% of the fatal errors were due to "performance and knowledge deficits," and 16% could be attributed to "communication errors" (FDA, 2011b). Medication labels if efficiently designed have the potential to significantly bridge the perceived performance, knowledge and communication gap. Because allowances have not been made for changes caused by aging, older adults are especially at risk. In the course of this study, it was found that there is a lot of information about health literacy, aging and other aspects of health care, but very limited data collected about the interaction of older adults and medication labels. Existing data suggests that there is much that can be done to increase the quality of life for older adults; such as modifying medication labels to suit the needs of the older adults. Given that a significant population of the patients in the U.S. in the next few years would be above 65 years old (about 56 million by 2030), there is a need to empower this patient group and minimize the potential financial burden associated with health care. The vulnerabilities of this patient group and the projected geriatric increase in this population, necessitate more research in this area.

REFERENCES

AHRQ and CDC. *2002.* Physical Activity and Older Americans: Benefits and Strategies. *Agency for Healthcare Research and Quality and the Centers for Disease Control.* Accessed February 4, 2012, http://www.ahrq.gov/ppip/activity.htm.

Anderson, J.L.,S. Dodman, and M. Kopelman, et al. 1979. Patient information recall in a rheumatology clinic. *Rheumatol Rehabil* 18: 245 -255.

AoA. 2011. Aging Statistics. *Administration on Aging U.S. Department of Health and Human Services* Accessed February 4, 2012. http://www.aoa.gov/aoaroot/aging_statistics/index.aspx

Bailey, S. C., W. Shrank, and R. M. Parker, et al. 2009. Medication label improvement: An issue at the intersection of health literacy and patient safety. *Journal of Communication in Healthcare* 2: 294-307.

Caban, A. L., D. J Lee, and O. Gómez-Marín, et al. 2005. Prevalence of concurrent hearing and visual impairment in US adults: the national health interview survey, 1997-2002. *Am J Public Health*. 95: 1940-1942.
CDC and NCCDPHP. 2011. Healthy Aging: Helping People to Live long and Productive Lives and Enjoy a Good Quality of Life. *Centers for Disease Control and Prevention National Center for Chronic Disease Prevention and Health Promotion.* Accessed February 3, 2012. http://www.cdc.gov/chronicdisease/resources/publications/AAG/aging.htm
Chermak, T. E. and B. L. Lambert. 2009. Descriptive analysis of primary package labels from commercially available prescription solid oral dosage form drugs. *Journal of the American Pharmacists Association* 49:399-406.
Coleman, E. A., E. H. Wagner, and L.C. Grothaus, et al. 1998. Predicting hospitalization and functional decline in older health plan enrollees: are administrative data as accurate as self-report? *J Am Geriatr Soc*. 46: 419-425.
Coppard, B.M, K. Coover, and M Faulkner. 2011. Use of Medication by Elders. *Occupational Therapy with Elders: Strategies for the COTA*. eds. R. Padilla, S. Byers-Connon and H. Lohman.
Dey, A. N. 1997. Characteristics of older nursing home residents: data from the 1995 National Nursing Home Survey. NCHS Advance Data 289:1-12.
Eaton, M.L., and R.L. Holloway. 1980. Patient comprehension of written drug information. *Am J Hosp Pharm* 37: 240-243.
FDA. 2011a. "Medication Errors." Food and Drug Administration: Protecting and Promoting Your Health. Accessed February 1, 2012. http://www.fda.gov/drugs/drugsafety/medicationerrors/default.htm
FDA, 2011b. "Strategies to Reduce Medication Errors: Working to Improve Medication Safety." Food and Drug Administration: Protecting and Promoting Your Health. Accessed February 18, 2012. http://www.fda.gov/Drugs/ResourcesForYou/Consumers/ucm143553.htm
Gazmararian J. A., D. W. Baker, M.V. Williams, et al. 1999. Health literacy among Medicare enrollees in a managed care organization. *JAMA*. 281: 545-551.
Ghetti, S. and P. J. Bauer. 2011. *Origins and Development of Recollection: Perspectives from Psychology and Neuroscience,* New York: Oxford University Press.
Hastings, S. N., J. C. Kosmoski and J. M. Moss. 2010. Special Considerations of Adherence in Older Adults. *Improving Patient Treatment Adherence: A Clinician's Guide*. ed. H. Bosworth. New York: Springer.
Horne R. 2006. Compliance, adherence, and concordance: implications for asthma treatment. *Chest* 130: 65S-72S.
Korsch, B., E. Gozzi, and V. Francis. 1968. Gaps in Doctor-Patient Communication: Doctor Patient Interaction and Patient Satisfaction. *Pediatrics* 42:855-871.
Kressels, R. P. C. 2003. Patients' memory for medical information. *Journal of the Royal Society of Medicine* 96: 219-222.
Kutner, M., E. Greenberg, Y. Jin, C. Paulsen. 2006. The health literacy of America's adults: Results from the 2003 National Assessment of Adult Literacy (NCES 2006-483). Washington D.C: National Center for Education Statistics, US Dept of Education.
Leonard, R. 2002. Statistics on Vision Impairment: A Resource Manual. *Arlene R. Gordon Research Institute of Lighthouse International.* 5
Marek, K.D., and L. Antle. 2008. Medication Management of the Community Dwelling Older Adult. In. *Patient Safety and Quality: An Evidence-Based Handbook for Nurses,*

ed. R.G. Hughes. Agency for Healthcare Research and Quality, Rockville, MD. Accessed January 17, 2012,*http://www.ahrq.gov/qual/nurseshdbk/*

McGuire, L.C.1996 Remembering what the doctor said: organization and older adults' memory for medical information. *Exp Aging Res* 22:403 -*428.*

Meadows, M. 2003. Strategies to Reduce Medication Errors: How the FDA Is Working to Improve Medication Safety and What You Can Do to Help. *FDA Consumer Magazine* 37:20-27

Ong, L. M. L., J. C. J. M. de Haes, and A. M. Hoos, et al. 1995. Doctor-patient communication: A review of the literature. *Social Science & Medicine* 40: 903-918

Parker, R. M. 2000. Health literacy: a challenge for American patients and their health care providers. *Health Promotion International* 15: 277–291.

Pignone, M., D. A. DeWalt, and S. Sheridan et al. 2005. Interventions to Improve Health Outcomes for Patients with Low Literacy: A systematic Review. *Journal of General Internal Medicine* 20: 185-192.

Quirk, P. A. 2000. Screening for Literacy and Readability: Implications for the Advanced Practice Nurse. *Clinical Nurse Specialist* 14: 26-32.

Rudd R. E., B. A. Moeykens, and T. A. Colton. 1999. Health and Literacy: A Review of Medical and Public Health Literature. In. *Review of Adult Learning and Literacy*. NCSALL. Vol 1. Lawrence Erlbaum Associates, Inc.

Shrank, W. H. and J. Avorn. 2007. Educating Patients About Their Medications: The Potential And Limitations Of Written Drug Information. *Health Affairs,* 26: 731-740.

Shrank, W. H, R. Parker, and T. Davis, et al. 2010. Rationale and design of a randomized trial to evaluate an evidence-based prescription drug label on actual medication use. *Contemp Clin Trials* 31: 564-571.

Smither, J. A. and C.C. Braun.1994. Readability of prescription drug labels by older and younger adults. *Journal of Clinical Psychology in Medical Settings* 1: 149-159.

Stoppler, M. C., and J.W. Marks. "The Most Common Medication Errors." Accessed February 18, 2012. http://www.medicinenet.com/script/main/art.asp?articlekey=55234

Taplay, K., and A. Flores-Vela. 2012. The Older Adult. Accessed January 29, 2012. http://www.wadsworthmedia.com/marketing/sample_chapters/9781111640460_ch13.pdf

Vredenburgh, A. G., and I. B. Zackowitz. 2009. Drug labeling and its impact on patient safety. *Work: A Journal of Prevention, Assessment and Rehabilitation* 33: 169-174.

Zuccollo, G. and H. Liddell. 1985. The elderly and the medication label: doing it better. *Age Ageing* 14: 371-376.

CHAPTER 23

Designing Virtual Reality Systems for Procedural Task Training

Nirit Gavish

Department of Industrial Engineering and Management,
Ort Braude College
Karmiel, Israel
nirit@braude.ac.il

ABSTRACT

The past decade has seen increasing use of virtual simulators for operators who need to acquire or practice procedural skills. As part of the effort to close the gap between technological advances in simulator capabilities and theoretical advances in training research, three experimental studies were conducted in order to formulate design guidelines for procedural task training in virtual reality systems. In all three studies, the ultimate goal of the training was assembling by hand a real LEGO® helicopter model requiring 75 steps. Study 1 demonstrated that partly observational learning can enhance training efficiency for procedural tasks without necessarily sacrificing performance if integrated properly within virtual reality training. A closer look at the contribution of cognitive training can be found in Study 2, which demonstrated that the two approaches to virtual training, physical fidelity and cognitive training methods, have complementary advantages. However, cognitive training should be used carefully, as demonstrated in Study 3, since providing enhanced information might make the active integration process more complex.

Keywords: Procedural Skills, Training, Virtual Reality, Cognition, Observational Learning

1 INTRODUCTION

Procedural knowledge, or procedural skill, is defined as knowing how and when (in what order) to execute a series of procedures needed to accomplish a certain task (Annett, 1996; Rittle-Johnson, Siegler, and Alibali, 2001). In procedural learning, the acquisition of cognitive skills moves from the declarative stage, in which instructions about the task are encoded as a set of facts that can be used by general interpretive procedures to generate behavior, to the procedural stage, in which knowledge is converted into a form that is directly applied without the intercession of other interpretive procedures; this process is achieved through repeated practice on the task (Anderson, 1982). Declarative knowledge corresponds to *knowing that*: knowledge about objects and processes that can be expressed in words. Procedural knowledge corresponds to *knowing how*: knowledge that enables people to interact effectively with the world even without explicit representation of that knowledge (Bucciarelli, 2007).

The past decade has seen increasing use of virtual simulators for operators who need to acquire or practice procedural skills – for instance, in the fields of laparoscopic surgery (Dawson, 2006), bronchoscopy (Colt, Crawford, and Galbraith, 2001), and industry and maintenance (Claessens, Min, and Moonen, 2000; Johnson and Rickel, 1997). However, as Salas, Bowers, and Rhodenizer (1998) pointed out, to enhance the effectiveness of virtual reality training, the focus should be not on designing technologically complex and realistic simulators, but on designing them based on learning needs. Similarly, Ward, Williams, and Hancock (2006) suggest that training has often been based upon historical precedents or fashionable constructs with only limited empirical evidence in support of their utility.

As part of the effort to close the gap between technological advances in simulator capabilities and theoretical advances in training research, three experimental studies were conducted in order to formulate design guidelines for procedural task training in virtual reality systems. Three main issues were addressed in these studies: integrating partly observational learning into virtual reality training systems; physical fidelity versus cognitive fidelity representation of the task; and the influence of providing enriched information during procedural task training on the development of learning patterns. In all three studies, the ultimate goal of the training was assembling by hand a real LEGO® helicopter model requiring 75 steps (see Figure 1). Completing this task was done with the assistance of an instruction manual with step-by-step diagrams of the 75 assembly stages, where each diagram includes a picture of the LEGO® bricks needed for that stage and the current view of the model (see Figure 2). For additional information about the studies, see Gavish, Gutierrez, Webel, Rodriguez, and Techia (2011); Gutierrez, Gavish, Webel, Rodriguez, and Tecchia (*in press*); Hochmitz and Yuviler-Gavish (2011).

Figure 1 The real 75-step LEGO® assembly task.

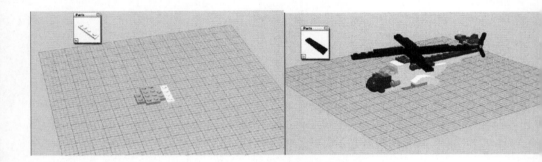

Figure 2 Instruction manual: step 3 (left) and step 75 (right).

2 EXPERIMENTAL STUDIES

2.1 Study 1: Integrating Partly Observational Learning

Sophisticated virtual reality training systems allow the trainee to enact the movements required to complete a task (e.g., Mellet-d'Huart, 2006; Stoffregen, Bardy, and Mantel, 2006). According to the enactive approach, training using a virtual reality training system is likely to be most effective if it is accompanied by physical actions. However, while the enactive approach for training is recommended for virtual reality training systems, it is also time-consuming, since the trainee needs to fully perform all the cognitive and physical activities involved in the task. Performing some of these activities may be necessary to acquire competence in the task, but it is possible that some of them could be successfully replaced with observational learning. Observational learning, in which a person watches the actions of another and either imitates or mentally records them, has been shown to be effective in training complex skills (McCullagh, Weiss, and Ross, 1989; Wouters, Tabber, and Pass, 2007; Wulf and Shea, 2002).

Study 1 investigate whether it is possible to utilize partly observational learning in order to improve training efficiency without necessarily sacrificing performance in virtual reality training systems: letting the trainee observe a portion of the task, which would be performed by the virtual reality system. This approach could be used to reduce training time, because the virtual system can be designed to perform faster than the human trainee – within the trainee's observational capacity but faster than his or her performance capability. Training efficiency would be improved if the resulting performance is not impaired.

The training platform in study 1 was a multimodal system comprised of a screen and a haptic device (see Figure 3). The 3-D graphical scene included a vertical back wall showing the set of the Lego bricks to be assembled, and a horizontal assembly space on which trainees would assemble the Lego model (see Figure 4). Trainees used the haptic device to grasp the pieces from the back wall and place them in their correct location in the assembly space. Two training conditions were employed in a between-participants design. In the Active condition, in each of the 75 steps a copy of the target brick to be assembled was first shown in its final position in the model. Trainees were then required to identify the target brick and move it to its correct position. In the Partly Observational Learning condition, the system provided the same information – that is, a copy of the target brick in its final position – but after the trainee selected the correct brick, the system automatically positioned it where it needed to go. A test in which participants assembled the real Lego structure using a printed instruction manual followed the training phase.

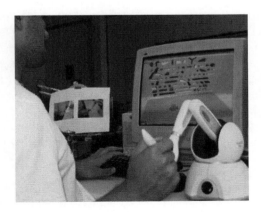

Figure 3 The virtual reality assembly training system.

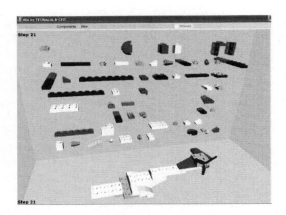

Figure 4 The virtual assembly scene.

The results demonstrated that mean performance times with the real Lego test were similar in both conditions, as well as the mean number of final errors, while mean training time for the Partly Observational Learning condition was significantly lower than that for the Active training. The mean number of corrected errors was somewhat lower in the Active condition, although not significantly. Hence, it seems that partly observational learning can enhance training efficiency for procedural tasks without necessarily sacrificing performance if integrated properly within virtual reality training.

2.2   Study 2: Physical Fidelity versus Cognitive Fidelity

Most researchers agree that procedural skill develops as a result of practice through repeated exposure to a certain task (e.g., Gupta and Cohen, 2002). Hence, ensuring that a training simulation resembles the actual task is likely to be important for successful transfer of training. There are two general approaches to ensure the resemblance between the training simulator and the actual task. The first approach emphasizes the importance of physical fidelity, defined as the degree to which the simulation looks, sounds, and feels like the actual task (see Alexander, Brunyé, Sidman, and Weil, 2005). The second approach holds that the simulator should incorporate high cognitive fidelity (Lathan, Tracey, Sebrechts, Clawson, and Higgins, 2002) with regard to the actual task. Cognitive fidelity refers to the extent that the simulator engages the trainee in the types of cognitive activities involved in the real-world task (Kaiser and Schroeder, 2003).

In Study 2, real-world training was compared to two alternative virtual trainers, one based on physical and the other on cognitive fidelity, for procedural skill acquisition in the LEGO® assembly task. Four training sessions with the virtual trainers were followed by a test in which participants constructed the actual model. One of the virtual trainers employed a 3-D manipulation environment which allowed trainees to perform the steps virtually but without haptic feedback; it thus emphasized physical fidelity of the task. The other required participants to verbally describe each step and the parts involved based on a diagram (a variant of motor imagery), and so emphasized cognitive fidelity of the task. The two virtual training methods were compared to each other and to real-world training based on participants' performance over the training sessions and in the real-world test. A control group performed the task without any training.

Overall, both the physical fidelity and cognitive fidelity training methods produced better performance time than no training at all, as did the real-world training. This finding thus confirms that both kinds of simulators can be useful for procedural skills training, especially at times when real-life training is impractical – for instance, when a device to be used is easily damaged, costly, or not always available, or when there is a possibility of risk to the person training for the task. However, a closer look at the results shows that the two virtual trainers were not equal in their contribution to skill transfer. The cognitive fidelity group was inferior in terms of test time compared with the real-world training, while the physical fidelity group fell somewhere in the middle, with no significant difference either way. In contrast, only the real-world and the cognitive fidelity groups, and not the physical fidelity group, had an advantage over the control group in terms of time allocated to error correction. These results suggest that the two approaches to virtual training have complementary advantages.

2.3   Study 3: Learning Patterns with Enriched Information

Although the emphasis in procedural skills is on *knowing how*, several studies have demonstrated that performance in a new procedural skill is more accurate, faster, and more flexible when learners are provided with enhanced information: specifically, "how-it-works" knowledge (Fein, Olson, and Olson, 1993; Kieras, 1988; Kieras and Bovair, 1984) in addition to "how-to-do-it" knowledge – or, in alternative terminology, "context procedures" in addition to "list procedures" (Taatgen, Huss, Dickison, and Anderson, 2008). However, under some conditions enhanced information does not improve procedural skills acquisition, for example, when the enhanced information does not allow learners to directly and simply infer the precise procedural steps required (Kieras and Bovair, 1984; Taatgen et al., 2008).

Study 3 focused on learning patterns developed with and without enriched information for procedural skill acquisition in the LEGO® assembly task. The preliminary computerized training application for the task presented the step-by-step assembly instructions from the manual, along with a short task to be performed at each step (selecting the relevant brick and pointing to its position in the model; see Figure 5). Trainees had 19 minutes to train with the computer application, after which they were required to build the real LEGO® model in the test phase. A Control group was trained with the computer application only, while participants in the experimental group (the Model group) were given the final assembled model to look at and touch while they trained.

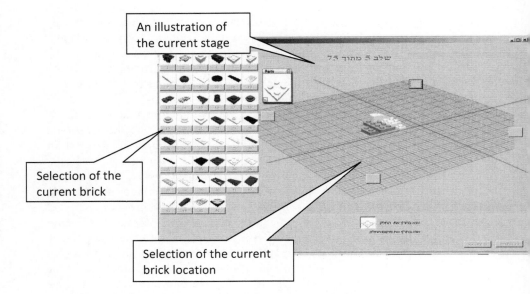

Figure 5 A screenshot of a diagram from the computerized training application.

While performance measures were similar between the two groups, their learning patterns were found to differ. Generally speaking, the Control group spent more of the training time in very short sessions, moving quickly from one diagram to another, while the Model group spent more time in long sessions. In addition, several learning patterns were correlated with poor performance in the Model group, namely overconfidence and over-reliance on the real model, but no correlations were found for the Control group. It seems that when the real model was provided during training, the active integration process became more complex.

3 CONCLUSIONS

Conclusions from this work may be implemented for the design of virtual reality training systems for procedural skills acquisition. Study 1 demonstrated that partly observational learning can enhance training efficiency for procedural tasks without necessarily sacrificing performance if integrated properly within virtual reality training. However, designers of virtual reality training systems must ensure that by incorporating observational learning, they do not make the training too easy and thereby inhibit mental processing (see Schmidt and Bjork, 1992). A closer look at the contribution of cognitive training in general can be found in Study 2, which demonstrated that the two approaches to virtual training, physical fidelity and cognitive training methods, have complementary advantages, and hence recom-mended that effective non-motor training for the development of procedural skills in psychomotor tasks should incorporate both physical fidelity and cognitive fidelity training.

However, cognitive training should be used carefully, as demonstrated in Study 3. Providing the real-world model during training made the active integration process (see Brunýe, Taylor, and Rapp, 2008; Mayer, 1984; Sternberg, 1985) more complex. The process complexity made it harder for participants to monitor their training and to choose the most effective training strategy. The improvement in learning can be achieved, possibly, if the trainee is aware of the metacognitive hazards of training under such circumstances. Directing trainees to be aware of overconfidence or to balance their time between the enriched information and the procedures might ultimately prove fruitful in terms of improving the design of procedural training applications. More research is needed to address these important abovementioned issues in the design of virtual reality training systems for procedural skills acquisition.

ACKNOWLEDGMENTS

This research was supported in part by the European Commission Integrated Project IP-SKILLS-35005. The Author wishes to thank all the scientists and technicians who were involved in the studies from DLR, FhG, CEIT, PERCRO, SIDEL, TECHNION and TECNALIA.

REFERENCES

Alexander, A. L., T. Brunýe, J. Sidman, and S. A. Weil. 2005. From gaming to training: A review of studies on fidelity, immersion, presence, and buy-in and their effects on transfer in pc-based simulations and games. *DARWARS Technical Report*.

Anderson, J. R. (1982). Acquisition of Cognitive Skill. *Psychological Review* 89: 369-406.

Annett, J. 1996. On knowing how to do things: a theory of motor imagery. *Cognitive Brain Research* 3: 65-69.

Brunýe, T. T., H. A. Taylor, and D. N. Rapp. 2008. Repetition and dual coding in procedural multimedia presentations. *Applied Cognitive Psychology* 22: 877-895.

Bucciarelli, M. 2007. How the construction of mental models improves learning. *Mind & Society* 6: 67–89.

Claessens, M., R. Min, and J. Moonen. "The use of virtual models for training procedural tasks". Paper presented at Workshop on Advanced Learning Technologies, IAWLT 2000, Palmerston North, New Zealand, 2000.

Colt, H. G., S. W. Crawford, and O. Galbraith, O. 2001. Virtual reality bronchoscopy simulation: A revolution in procedural training. *Chest* 120: 1333-1339.

Dawson, M. 2006. Procedural simulation: A primer. *Radiology* 241: 17-25.

Fein, R. M., G. M., Olson, and J. S. Olson. "A mental model can help with learning to operate a complex device". Paper presented at Conference on Human Factors in Computing Systems, Amsterdam, the Netherlands, 1993.

Gavish, N., T. Gutierrez, S. Webel, J. Rodriguez, and F. Techia. "Design guidelines for the development of virtual reality and augmented reality training systems for maintenance and assembly tasks". Paper presented at The International Conference of the European SKILLS Project, Montpellier, France, 2011.

Gupta, P. and N. J. Cohen. 2002. Theoretical and Computational Analysis of Skill Learning, Repetition Priming, and Procedural Memory. *Psychological Review* 109: 401-448.

Gutierrez, T., N. Gavish, S. Webel, J, Rodriguez, and F. Tecchia, F. In press. Training platforms for Industrial Maintenance and Assembly. In *Skills training in multimodal virtual environments*, eds. M. Bergamasco, B. G. Bardy, and D. Gopher. Taylor & Francis.

Hochmitz, I. and N. Yuviler-Gavish, N. 2011. Physical fidelity versus cognitive fidelity training in procedural skills acquisition. *Human Factors* 53: 489-501.

Johnson, W. L. and J. Rickel. 1997. Steve: An animated pedagogical agent for procedural training in virtual environments. *ACM SIGART Bulletin* 8: 16-21.

Kaiser, M. K., and J. A. Schroeder. 2003. Flights of fancy: The art and science of flight simulation. In *principles and practice of aviation psychology*, eds. P. M. Tsang and M. A. Vidulich. Lawrence Erlbaum: NJ.

Kieras, D. E. 1988. What mental model should be thought: Choosing instructional content for complex engineered systems. In *Intelligent tutoring systems: Lessons learned*, eds. J. Psotka, L. D. Massey, and S. A. Mutter. Hillsdale, NJ: Lawrence Erlbaum Associates.

Kieras, D. E. and S. Bovair. 1984. The role of mental model in learning to operate a device. *Cognitive Science* 8: 255-273.

Lathan, C. E., M. R. Tracey, M. M. Sebrechts, D. M. Clawson, and G. A. Higgins. 2002. Using virtual environments as training simulators: Measuring transfer. In *Handbook of virtual environments: Design, implementation, and applications*, ed. K. M. Stanney. Mahwah, NJ, US: Lawrence Erlbaum Associates.

Mayer, R. E. 1984. Aids to prose comprehension. *Educational Psychologist* 19: 30-42.

McCullagh, P., M. R.Weiss, and D. Ross. 1989. Modeling considerations in motor skill and performance: An integrated approach . In *Exercise and sports science seview* (Vol. 17), ed. K. Pandolf. Baltimore: Williams & Wadkins.

Mellet-d'Huart, D. 2006. A model of (en)action to approach embodiment: A cornerstone for the design of virtual environment for learning. *Virtual Reality* 10: 253-269.

Rittle-Johnson, B., R. Siegler, and M. Alibali. 2001. Developing conceptual understanding and procedural skill in mathematics: An iterative process. *Journal of Educational Psychology* 93: 346–362.

Salas, E., C. A. Bowers, and L. Rhodenizer. 1998. It is not how much you have but how you use it: Toward a rational use of simulation to support aviation training. *International Journal of Aviation Psychology* 8: 197-208.

Schmidt, K. and D. Bjork. 1992 New conceptualization of practice: Common principles in three paradigms suggest new concepts for training. *Psychological Science* 3: 207-217.

Sternberg, D. A. 1985. *Beyond IQ: A triarchic theory of human intelligence*. Cambridge, England: Cambridge University Press.

Taatgen, N. A., D. Huss, D. Dickison, and J. R. Anderson. 2008. The acquisition of robust and flexible cognitive skills. *Journal of Experimental Psychology: General* 137: 548-565.

Ward, P., A. M. Williams, and P. A. Hancock. 2006. Simulation for performance and training. In *Cambridge handbook of expertise and expert performance*, eds. K.A. Ericsson, N. Charness, R. Hoffman, and P. Feltovich. Cambridge: Cambridge University Press.

Wouters, P., H. K. Tabbers, and F. Pass, F. 2007. Interactivity in video-based models. *Educational Psychology Review* 19: 327-342.

Wulf, G. and C.H. Shea. 2002. Principles derived from the study of simple skills do not generalize to complex skill learning. *Psychonomic Bulletin & Review* 9: 185-211.

Section VI

Neuroergonomics: Measurement

CHAPTER 24

Given the State, What's the Metric?

Meredith Carroll, Kay Stanney, Christina Kokini, Kelly Hale

Design Interactive, Inc.
Oviedo, FL
meredith@ kay@ christina@ kelly@designinteractive.net

ABSTRACT

Today's resource constrained environment has pushed the training community toward more intelligent, automated training systems, which provide an opportunity to reduce instructor requirements and increase training throughput. One challenge with this, however, is that in order for an intelligent training system to achieve individually tailored training that is similar to the capability of a human instructor, the system must be able to assess trainee performance, and how that performance is influenced by a trainee's affective (e.g., frustration, anxiety) and cognitive (e.g., workload, engagement) learning states. With the emergence of advanced measurement technology (e.g., eye tracking, electroencephalography, etc.) access to these cognitive and affective states by an intelligent automated system is becoming more feasible. This paper reviews behavioral, cognitive, and physiological metrics that have been validated to provide an indication of learning states and thus could be used to derive the diagnostics necessary to drive tailored training solutions.

Keywords: affective state, cognitive state, adaptive training, human performance assessment, human performance metrics, training effectiveness and efficiency

1 INTRODUCTION

It's simple... If you cannot measure human performance, you cannot optimize it. If you think about it, it seems downright irresponsible that we still often make training system acquisition decisions with a thumbs-up or -down from decision makers. Without understanding if a given solution will optimize human performance, we adopt it because people "think" it will work. Perhaps that's why we embrace the term "human-systems integration" rather than "human-systems engineering" – so we can get away with "soft" science practices, rather than an engineering approach that relies on basic scientific and mathematic principles to develop useful human-system interactive tools and objects. It is time we demanded more – more hard science, more algorithms characterizing human performance, and especially - more measurement. There was a time when this would have been impossible, as there was no way to delve inside the mind of a trainee and identify just how effective a training system was at fostering learning. This roadblock no longer exists, as there are now ample ways to measure the dimensions of human performance – psychomotor, cognitive, and affective - and a growing number of ways to transform these data into useful information to enhance training effectiveness and efficiency. This information can, in turn, be used to develop intelligent training systems that tailor training to meet the needs of an individual trainee, in a manner similar to the capability of a (good!) human instructor. Much like a human instructor, the system would be able to detect, for example, the affective state of frustration, which could be experienced during a cognitive state of overload, thus indicating the need to slow the pace of instruction. The result could be instruction that is as seamless and intuitive as that achieved by the most brilliant of instructors. This paper focuses on cognitive and affective learner states. It reviews those states that strongly influence learning, discusses how these states impact learning, summarizes ways to measure these states, and sheds light on the advantages and disadvantages of various measurement approaches.

2   WHAT'S THE STATE?

Many cognitive and affective states influence the ability to learn. This section will review those states that have been found to be most influential in learning.

2.1   Cognitive States

There are several cognitive states that impact the readiness of a trainee to learn, such as attention, engagement, distraction, drowsiness, and workload. Each of these states, when not within optimal levels, can reduce learning and retention. However, when optimized, these states can maximize learning opportunities.

Attention refers to the cognitive process which gives rise to conscious awareness by selecting between competing stimuli (Naish, 2005). Attention can result either from one directing their attention to relevant cues in an environment based on a priori knowledge or from a salient stimulus in the environment directing attention to itself (i.e., bottom up processing). High levels of attention appear to be critical during learning, as reduced attention levels have been shown to result in lower retention and less ability to apply information (Small, Dodge & Jiang, 1996; McQuiggan, Lee, and Lester, 2007). Further, divided attention during encoding has led to significant reductions in recall (Craik et al., 1996). Thus, optimization of attention during learning has potential to increase retention and recall capabilities.

Engagement can be defined as the level of cognitive processes related to information gathering and sensory processing (Berka et al., 2007). High engagement reflects attentional focus, whereas low levels of engagement indicate that a trainee is not actively engaged with some aspect of the environment (Dorneich et al., 2004). Disengagement has been show to be negatively correlated with learning (Woolf, Burleson, and Arroyo, 2007) and performance gains (Johns and Woolf, 2006). Studies have also shown a trend of decreasing levels of engagement with increasing task proficiency (Berka et al., 2007; Stevens, Galloway, & Berka, 2007). Together, these findings suggest that high levels of engagement are critical during early stages of learning, however, as expertise increases, it is anticipated that high levels of engagement are less critical.

Distraction is a state characterized by a lack of clear and orderly thought and behavior, where a learner becomes involved somewhere other than the cognitive tasks of interest (Poythress et al., 2006). Learning can be inhibited when a trainee is distracted; even if it does not decrease the overall level of learning, it may result in acquisition of knowledge that is less flexibly applied in new situations (Foerde, Knowlton, & Poldrack, 2006). This suggests it is critical to minimize learner distraction to increase skill acquisition and transfer.

Drowsiness can result from sleep disorders, which are common and can have deleterious effects on performance (Berka et al., 2005). Loss of sleep can accumulate over time and result in a "sleep debt," which can lead to impairments in alertness, memory, and decision-making. Such fatigue produces lapses in attention that have been measured in continuous performance paradigms (c.f. Rosekind, Neri, & Dinges, 1997). Further, memory has been shown to lead to self-report of memory deficits (Friedman et al., 1971), a key player in learning and retention. Thus, it is key to minimize learner drowsiness to ensure adequate levels of attention and memory to facilitate effective learning.

Workload has been described as the relation between the mental resources demanded by a task and those resources available to be supplied by the human operator (Parasuraman, Sheridan, & Wickens, 2008). High levels of workload can cause delays in information processing or users to ignore or misinterpret incoming information (Ryu & Myung, 2005). However, minimizing workload is not necessarily ideal as workload has been shown to have an inverted-U relationship with both motivation and performance for low to medium levels of expertise, wherein, these constructs were maximized at moderate workload levels (Bendoly & Prietula, 2008). Further, mental workload has been shown to mediate human performance of perceptual and cognitive tasks (Parasuraman & Caggiano, 2002), both critical processes engaged during the learning process, thus it is necessary to ensure that workload is optimized to maximize learning opportunities.

2.2 Affective States

Literature has established that affective states such as anger, fear, boredom, confusion, and confidence can influence the learning process, sometimes supporting and sometimes impeding learning and retention (Small et al., 1996; Burleson and Picard, 2004; Woolf et al., 2009). These affective states are generally characterized by two dimensions - valence (positive-negative nature of the emotion) and arousal (calming/soothing to exciting/agitating; (Baker et al., 2010; Russell, 2003).

Anger is a negative, high arousal emotion that occurs at the prospect of a negative event with consequences to oneself (Ortony, et al., 1988). Frustration is a term often used in the context of learning; it is a different intensity level of the same affective state as anger (Scherer, 2005). Frustration tends to impede learning, as when learners become frustrated they don't take in information efficiently (Woolf et al., 2009; Burleson and Picard, 2004). This may be because they concentrate on the source of the anger, thereby diverting their attention from learning (McQuiggan, Lee, & Lester, 2007).

Fear and anxiety, displeasing feelings of concern that can result in worry, uneasiness, and dread, represent different intensity levels of the same affective state (Scherer, 2005). Anxiety has been found to have negative effects on academic achievement (Pekrun, 1992; 2002); similar to frustration, anxiety can cause a learner to concentrate on the source of the anxiety, distracting from the task at hand, thereby impeding learning (McQuiggan, Lee, & Lester, 2007).

Boredom, a negative, low arousal emotion that occurs when a situation is construed to be monotonous or dull (Merrifield, 2010), is often negatively correlated with learning gains (Craig et al., 2004). Boredom also leads to lower retention and less ability to apply information in the future (Small et al., 1996). Baker et al. (2010) found that once a student was bored it was difficult to transition out of that state, indicating that it is important to prevent boredom in the first place.

Confusion is a negative, medium arousal emotion, which occurs when a learner reaches an impasse that provides an opportunity for learning. It is beneficial to learning in short-term instances, but becomes detrimental when experienced for long periods of time (Baker et al., 2010). Craig et al. (2004) found that learners who were confused during a learning session performed better than those who didn't; suggesting that some level of confusion may be necessary for optimal learning. Further, learners who are confused may be less likely to become disengaged and transition to the state of boredom (D'Mello et al., 2007).

Confidence is a positive emotion with a medium level of arousal (Scherer, 2005), and is the degree to which a person feels certain of his or her abilities (Hart, 1989). Norman and Hyland (2003) found that increasing learners' confidence allowed them to engage more fully in the learning process and increased motivation.

Table 1 below provides a summary of the cognitive and affective states discussed above and their impact on learning.

Table 1 Cognitive and Affective States that Impact Learning

State	Impact on Learning
Attention	High levels of attention during learning can increase retention/recall.
Engagement	High levels of engagement are critical during early stages of learning.
Distraction	Minimize learner distraction to increase skill acquisition and transfer.
Drowsiness	Minimize learner drowsiness to facilitate effective learning.
Workload	Ensure workload is optimized (not too high/too low) to maximize learning.
Anger	Anger impedes learning by limiting efficient intake of information.
Fear	Fear distracts from the task at hand, thereby impeding learning.
Boredom	Boredom leads to lower retention and less ability to apply information.
Confusion	Confusion is detrimental when experienced for long periods of time.
Confidence	High levels of confidence facilitate more engaged learning.

3 WHAT'S THE METRIC?

Given the criticality of the above cognitive and affective states to learning, a key step in optimizing training is the ability to quantify such states. There are multiple ways to assess these states, including survey (knowledge-based and self-reports), behavioral, and physiological methods. This section will review some of the metrics available to assess these states and advantages and disadvantages of each.

3.1 Cognitive State Metrics

Traditionally, cognitive state measures have taken the form of surveys/paper-based assessments and behavioral performance assessments. While self-report measures provide insights into cognitive state, they rely on one's conscious awareness and recall of the state. Further, these measures are typically captured at discrete points in time, and cannot be reliably captured in real-time on a continuous basis, thus limiting the understanding of learner state over time. Knowledge-based questionnaires provide a more objective measure of state, but are still limited to discrete points in time. While behavioral measures are typically easy to capture, they do not allow for real time measurement. Observer-based behavioral measures are generally easy to define, but require trained observers to accurately and reliably evaluate. With the emergence of physiological monitoring technology, there has been a growing interest in utilizing physiological assessment techniques, as they can provide objective, non-obtrusive, real-time data regarding a variety of cognitive states. However, they require equipment, which can be expensive and physically invasive. Table 2 reviews survey, behavioral and physiological metrics of cognitive states; all sources can be found in Carroll, Stanney, Kokini and Hale (2012).

In terms of the strengths and weaknesses of specific cognitive state metrics, for visual attention, eye tracking can provide an objective, relatively precise measure but requires use of technology which can be cost prohibitive. For drowsiness, EOG and PERCLOS provide reliable, continuous fatigue measure, however, it may be best to combine them to achieve the highest reliable and validity. For workload, all surveys listed in the table have been shown to be sensitive to different levels of workload; however, the Workload Profile has been shown to have the highest levels of sensitivity and diagnosticity. Behavioral workload metrics, such as performance-based measures, cannot capture workload changes that are countered by changes in effort. Further, some secondary tasks may be insensitive to primary task demands.

3.2 Affective State Metrics

Traditionally, affective state measures have taken the form of self-report surveys. The strength of subjective methods include their ease of administering, and often times reliable results (Champney, 2009). Nonetheless, they are disruptive to the task/experience being assessed, rely on the individuals' memory, and ability to express their emotions, and rely on the honesty of individuals' reports. With the emergence of sensor technology, behavioral and physiological measures provide an opportunity to assess affective states in a more objective manner, with lower levels of task intrusion and higher levels of temporal resolution. These measures, however, require specialized equipment, which can be expensive and physically invasive. Table 3 reviews survey, behavioral and physiological metrics used to gauge affective states; all sources can be found in Carroll et al. (2012).

In terms of the specific strengths and weaknesses of affective state metrics, The EmoPro^{TM} survey measure allows assessment without the verbal bias existing in verbal-based instruments, however, is limited in temporal resolution and relies on awareness and recall. Postural changes can be readily identified by a trained observer without distraction of the subject of interest (Champney, 2009). However, such observer-based metrics are only valid for a small number of emotions.

4 CONCLUSIONS

This review provides a toolbox of cognitive and affective state metrics that can be used to drive individually tailored training that is similar to the capability of a human instructor. Such metrics can be used to assess trainee performance, and determine how that performance is influenced by a trainee's affective and cognitive learning states. These metrics provide powerful tools in understanding a user's performance. They allow one to peer into the mind of the user to get at the "Why?" behind performance errors. Keep in mind, however, that selecting the right metrics is context dependent and decisions will be driven by many factors, including measurement goals, cost, schedule, and performance constraints.

ACKNOWLEDGMENTS

This material is based upon work supported in part by ONR under Contract No. N00014-11-M-0101 and ARL under Contract No. WR1CRB-10-C-0327.

Table 2 Cognitive State Metrics

Cognitive State	Survey, Behavioral, and Physiological Metrics
Attention	**Survey Metrics:** Verbal recall of information presented textually
	Behavioral Metrics: Detection of event, as indicated by a behavioral response; Occurrence of and time spent viewing specific information; Eye tracking: Number of and duration of fixations, % of gaze time on Area of Interest (AOI)
	Physiological Metrics: EEG Gamma-Band Response (GBR); Gamma frequency oscillations
Engagement	**Survey Metrics:** User Engagement Questionnaire; User Engagement in Multimedia Presentation Survey; Motivation and Strategy Use Survey; Game Engagement Questionnaire
	Behavioral Metrics: Time on task; long term measure of cumulative engagement and more acute response using posture/position sensors on individual and pressure sensors within a chair
	Physiological Metrics: EEG engagement indices
Distraction	**Survey Metrics:** Self report
	Behavioral Metrics: Key press response time/accuracy related to target and distraction items; Primary task performance on 2-back secondary task
	Physiological Metrics: EEG distraction index
Drowsiness	**Survey Metrics:** Epworth Sleepiness Scale; Stanford Sleepiness Scale; Karolinska Sleepiness Scale; Visual analogue scales of sleepiness
	Behavioral Metrics: Psychomotor Vigilance Task; Percentage of Eye Closure (PERCLOS); Electrooculography (EOG): slow roving eye movement, reduced max saccadic velocity, eyelid reopening velocity
	Physiological Metrics: EEG activity in the alpha, theta, and delta bands, large alpha bursts, EEG models provide early warning of the onset of drowsiness
Workload	**Survey Metrics:** NASA Task Load Index (NASA TLX); Subjective Workload Assessment Technique (SWAT); Modified Cooper Harper Scale; Workload Profile (WP) Survey
	Behavioral Metrics: Primary task performance; Secondary task performance
	Physiological Metrics: EEG workload indices, oscillations in the alpha band, alpha peak, theta-band activity, pupil diameter, pupil amplitude variation

Table 3 Affective State Metrics

Affective State	Survey, Behavioral, and Physiological Metrics
Anger	**Survey Metrics:** State-Trait Anger Expression Inventory **Behavioral Metrics:** Posture change, Facial Expression, Vocal Pitch **Physiological Metrics:** Electromyography, Respiration rate, Heart rate, Finger temperature, Galvanic Skin Response (GSR)
Fear	**Survey Metrics:** EmoPro™ **Behavioral Metrics:** Vocal Pitch **Physiological Metrics:** Heart rate, GSR, Mean Pre-Ejection Period, Mean InterBeat Interval, Mean temperature, Mean pulse transit time, Sympathetic activity power
Boredom	**Survey Metrics:** Multidimensional State Boredom Scale, EmoPro™ **Behavioral Metrics:** Facial Expression, Vocal Pitch **Physiological Metrics:** Heart Rate, GSR
Confusion	**Survey Metrics:** EmoPro™ **Behavioral Metrics:** Posture change, Facial Expression, Vocal Pitch **Physiological Metrics:** Because the state of confusion has either neutral or variable levels of arousal, valid physiological metrics of these states have not been achieved.
Confidence	**Survey Metrics:** State Metacognitive Inventory, Self-esteem survey **Behavioral Metrics:** Facial Expression **Physiological Metrics:** Because the state of confidence has either neutral or variable levels of arousal, valid physiological metrics of these states have not been achieved.

REFERENCES

Baker, R., S. J., D'Mello, S. K. and Rodrigo, M. T., et al., 2010. Better to be frustrated than bored: The incidence, persistence, and impact of learners' cognitive–affective states during interactions with three different computer-based learning environments. *International Journal of Human-Computer Studies* 68: 223-241.

Bendoly, E. and Prietula, M. 2008. In "the zone", the role of evolving skill and transitional workload on motivation and realized performance in operational tasks. *International Journal of Operations and Production Management* 28 (12): 1130–1152.

Berka C, Levendowski DJ, and Lumicao MN, et al. 2007. EEG correlates of task engagement and mental workload in vigilance, learning, and memory tasks. *Aviation Space and Environmental Medicine* 78(5, Suppl.): B231-B244.

Berka, C., Levendowski, D.J., and Westbrook, P. et al. 2005. EEG quantification of alertness: Methods for early identification of individuals most susceptible to sleep deprivation. *Proceedings of the SPIE Defense and Security Symposium, Biomonitoring for Physiological and Cognitive Performance during Military Operations* 5797: 78-89.

Burleson, W. and Picard, R. "Affective Agents: Sustaining Motivation to Learn Through Failure and a State of 'Stuck'". Workshop on Social and Emotional Intelligence in Learning Environments. 7th International Conference on Intelligent Tutoring Systems, Maceio – Alagolas, Brasil, 2004.

Carroll, M., Stanney, K., and Kokini, C., et al. 2012. A review of cognitive and affective state measures. (Internal Technical Report). Oviedo, FL: Design Interactive.

Champney, R.C. 2009. The Integrated User Experience Evaluation Model: A Systematic Approach to Integrating User Experience Data Sources (Unpublished doctoral dissertation). University of Central Florida: Orlando.

Craig, S. D., Graesser, A. C., and Sullins, J., et al. 2004. Affect and learning: An exploratory look into the role of affect in learning with AutoTutor. *Journal of Educational Media* 29(3): 241-250.

Craik, F. I., Govoni, R., and Naveh-Benjamin, M., et al. 1996. The effects of divided attention on encoding and retrieval processes in human memory. *Journal of Experimental Psychology General* 152(2), 159-180.

D'Mello, S., Chipman, P., and Graesser, A. 2007. Posture as a predictor of learner's affective engagement. *Proceedings of the 29th Annual Meeting of the Cognitive Science Society* 905–910.

D'Mello, S. K., Taylor, R., and Graesser, A. C. (2007). Monitoring Affective Trajectories during Complex Learning. In D. S. McNamara and J. G. Trafton (Eds.), *Proceedings of the 29th Annual Cognitive Science Society* 203-208. Austin: Cognitive Science Society.

Dorneich, M. Whitlow, S. D., and Ververs, P. M. et al. 2004. DARPA Improving Warfighter Information Intake under Stress - Augmented Cognition Concept Validation Experiment (CVE) Analysis Report for the Honeywell Team. Prepared under contract: DAAD16-03-C-0054.

Foerde, K., Knowlton, B. J., and Poldrack, R. A. 2006. Modulation of competing memory systems by distraction. *Proceedings of the National Academy of Sciences of the United States of America* 103(31): 11778-11783.

Friedman, R. C., Bigger, T. J., and Kornfeld, D. S. 1971. The Intern and Sleep Loss. *New England Journal of Medicine* 285: 201-203.

Hart, L. E. 1989. Classroom Processes, Sex of Student, and Confidence in Learning Mathematics. *Journal for Research in Mathematics Education* 20(3): 242-260.

Johns, J. and Woolf, B. 2006. A Dynamic Mixture Model to Detect Student Motivation and Proficiency. *Proceedings of the Twenty-First National Conference on Artificial Intelligence* 2-8. Menlo Park, CA: AAAI Press.

McQuiggan, S., Lee, S., and Lester, J. 2007. Early prediction of student frustration. *Affective Computing and Intelligent Interaction* 698-709.

Merrifield, C. 2010. Characterizing the Psychophysiological Signature of Boredom. (Unpublished masters thesis). University of Waterloo, Waterloo.

Naish, P. 2004. Attention. In , N. Braisby & A. Gellatly (Eds.), *Cognitive Psychology*. Oxford University Press.

Norman, M. and Hyland, T. 2003. The Role of Confidence in Lifelong Learning. *Educational Studies* 29(2-3): 261-272.

Ortony, A., Clore, G. L., and Collins, A. 1988. *The Cognitive Structure of Emotions*. Cambridge: Cambridge University Press.

Parasuraman, R. and Caggiano, D. 2002. Mental workload, in V. S. Ramachandran (ed.), *Encyclopedia of the Human Brain,* Vol. 3. San Diego: Academic Press, 17–27.

Parasuraman, R., Sheridan, T. B, and Wickens, C.D. 2008. Situation Awareness, Mental Workload, and Trust in Automation: Viable, Empirically Supported Cognitive Engineering Constructs. *Journal of Cognitive Engineering and Decision Making* 2(2): 140-160.

Pekrun, R. (1992). Expectancy-value theory of anxiety: Overview and implications. D. G. Forgays, T. Sosnowski, & K. Wrzesniewski (Eds.), *Anxiety: Recent developments in self-appraisal, psychophysiological and health research* 23–41. Washington, DC: Hemisphere.

Pekrun, R., Goetz, T., and Titz, W., et al. 2002. Academic Emotions in Students' Self-Regulated Learning and Achievement: A Program of Qualitative and Quantitative Research. *Educational Psychologist* 37(2): 91-106.

Poythress, M., Russell, C., and Siegel, S., et al. 2006. Correlation between expected workload and EEG indices of cognitive workload and task engagement. In D. Schmorrow, K. Stanney, & L. Reeves (Eds.), *Augmented Cognition: Past, Present and Future* 32-44. Arlington, VA: Strategic Analysis, Inc.

Rosekind, M. R., Neri, D. F., and Dinges, D. F. 1997. From laboratory to flightdeck: Promoting operational alertness. *Fatigue and Duty Limitations? An International Review* 7.1– 7.14. London: The Royal Aeronautical Society.

Russell, J., 2003. Core affect and the psychological construction of emotion. *Psychological Review* 110: 145–172.

Ryu, K. and Myung, R. 2005. Evaluation of mental workload with a combined measure based on physiological indices during a dual task of tracking and mental arithmetic. *International Journal of Industrial Ergonomics* 35: 991-1009.

Scherer, K. R. (2005). What are emotions? And how can they be measured? *Social Science Information* 44(4): 695-729.

Small, R. V., Dodge, B. J., and Jiang, X. "Dimensions of interest and boredom in instructional situations." Proceedings of Selected Research and Development Presentations at the 18th National Convention of the Association for Educational Communications and Technology, Indianapolis, IN, 1996.

Stevens, R., Galloway, T., and Berka, C. "EEG-Related Changes in Cognitive Workload, Engagement and Distraction as Students Acquire Problem Solving Skills." User Modeling Conference, Athens, Greece, 2007.

Woolf, B., Burelson, W., and Arroyo, I. 2007. Emotional intelligence for computer tutors. *Supplementary Proceedings of the 13th International Conference of Artificial Intelligence in Education*, 6-15.

Woolf, B., Burleson, W., and Arroyo, I., et al. 2009. Affect-aware tutors: recognizing and responding to student affect, *International Journal of Learning Technology* 4(3/4): 129–164.

CHAPTER 25

Development of a Neuroergonomic Application to Evaluate Arousal

Daniel Gartenberg, Ryan McGarry, Dustin Pfannenstiel, Dean Cisler,

Tyler Shaw, Raja Parasuraman

George Mason University
Fairfax, VI 22031, USA

ABSTRACT

We developed and tested a neuroergonomic smartphone application called Mind Metrics that can be used to evaluate vigilance and working memory under naturalistic conditions. The application met a requirement to the field of neuroergonomics because the cognitive tasks were made for a smartphone platform, allowing the ability to make predictions about the neural processes that impact human performance during naturalistic work related activities (i.e. ubiquitous computing). However, if naturalistic tasks are to be developed that are sensitive to cognitive processes, these tasks must be tested and evaluated for validity by comparing performance to data obtained in controlled laboratory environments. In this study, we developed tasks that measure working memory and vigilance, two processes that are well known to affect human performance at work. We then tested participants on these tasks using both a smartphone and a desktop computer platform. The tasks we used to measure vigilance included a vigilance task called the Psychomotor Vigilance Task (PVT) and a vigilance task called the Spatial Discrimination Vigilance Task (SDVT). To measure working memory, we used a Color N-back Task (CNB) and a Spatial N-back Task (SNB). Using a mixed group design, participants were assigned to a desktop or smartphone condition and completed all four tasks. As predicted, there was a vigilance decrement for both the PVT and the SDVT, which was demonstrated by an overall slowing of responses as the tasks progressed. This decrement occurred for both the smartphone and the desktop tasks. Another interesting finding related to improvement over time for the N-back tasks-- participants performed faster on the n-back tasks as the task progressed. This indicates that task learning is an important factor to consider when developing neuroergonomic tasks aimed at detecting cognitive functioning in the wild. In previous research it was found that increased resource demands exacerbate the vigilance decrement. These findings suggest that learning can play a role in attenuating the vigilance decrement effect in tasks with high resource demands. If vigilance tasks developed on the smartphone can be administered in naturalistic environments this platform will provide a method of easy-to-obtain samples of repeated task performance, thereby reducing the impact of learning effects that can mask the vigilance decrement. The possible implications of this research are a more sensitive measure of the vigilance decrement for detecting vigilance in the wild.

Keywords: Neuroergonomics, Brain Arousal, Vigilance, Sleep, Health Care

1 INTRODUCTION

Through the merger of neuroscience, human factors psychology, and engineering, neuroergonomics aims to optimize mental functioning during cognitive and physical work (Parasuraman, 2003). The human brain's arousal system exerts an important influence on performance in work environments for both simple and complex tasks (Balkin, Rupp, Picchioni, et al., 2008). Lower arousal is associated with increased rate of accidents (Stutts, Wilkins, Osberg, et al., 2003). There is therefore a need to develop a neuroergonomic application that can provide the worker and co-worker with feedback on the worker's current level of alertness (Rizzo, Robinson, & Vicki, 2007). Yet there are obstacles to obtaining data in real-world settings because these settings require the use of different measurement tools. Additionally, the worker is less willing to devote a large amount of daily time to participate than participants in conventional laboratory experiments. At the same time there is the potential to obtain repeated measures over a long period of time when utilizing naturalistic data collection techniques.

Smartphones can be used as research tools to easily collect data in naturalistic environments; however, it is unclear how findings generalize across smartphone and desktop platforms and whether the former provide similar data to that obtained under controlled laboratory conditions. The iPhone, in addition to operating under different processing speeds, utilizes a touch screen interface and software that samples at different rates than desktop software that is typically used in laboratory experiments. Therefore, tasks requiring precise timing measurements, such as simple reaction time tests, may not produce the same results across platforms.

When developing a neuroergonomic application for the detection of variations in alertness in the workplace it is important to choose a test that is both sensitive and has minimal time costs. There are many laboratory-based tests and questionnaires that have been used to assess arousal (Matthews, Davies, Westerman, et al., 2000). While laboratory-based tasks, such as the psychomotor vigilance task (PVT) (Dinges and Powell, 1985), are usually more sensitive than questionnaires at detecting alertness changes (Van Dongen, Maislin, Millington, et al., 2003), laboratory-based tasks frequently require long periods of at least 10 minutes to administer (Dinges and Powell, 1985).

The ideal measure of alertness is therefore a task-based measure that can be administered for a short period of time. Based on the resource theory of vigilance (Parasuraman, 1985; Warm, Parasuraman, Matthews, 2008), a vigilance task that requires more resources will be more sensitive to a decrement in performance than one that requires less cognitive resources. Since resources are depleted during low arousal states (Dinges, Orne, Whitehouse, et al., 1987), a high resource demanding vigilance task will be more sensitive to variations in worker arousal. The longitudinal and repeated data collected on the smartphone can be used to address the learning effects that are likely to occur in these more resource demanding tasks.

2 VIGILANCE

Vigilance tasks typically involve the detection of signals over a long period of time, that are intermittent, unpredictable, and infrequent. As vigilance tasks progress, performance steadily declines, and there is a marked steep decline at about 20 minutes (Davies & Parasuraman, 1982; Parasuraman, 1986; Boksem, Meijman, Lorist, 2005, Lim, Wi, Wang, et al., 2010), although some evidence suggests that the decrement function is complete after 5 minutes in especially demanding circumstances (e.g. Helton, Dember, Warm, & Matthews, 2000). A conceptual framework for understanding the vigilance decrement performance is provided by resource theory (Parasuraman & Davies, 1984; Parasuraman, Warm, & Dember, 1987; Warm & Dember, 1998; Warm, Parasuraman & Matthews, 2008) within which the decrement results from the depletion of information-processing assets that cannot be replenished during continuous task performance.

Vigilance tasks require discriminations that either involve holding a representation in working memory (successive vigilance task) and comparing that representation with the current image, or the information needed to make the discrimination is presented on the screen, and no or very little working memory is required (simultaneous vigilance tasks) (Parasuraman, 1979). Given that successive tasks require working memory and simultaneous tasks only require comparative judgments, performance usually degrades more steeply and quickly in the more demanding successive task condition (e.g. Caggiano & Parasuraman, 2004)

The PVT is the gold standard task used by sleep researchers to measure the arousal system and it has been found to be sensitive to all the components of sleep (Dinges, et al., 1987; Van Dongen, et al., 2001). The PVT is a simultaneous vigilance task that requires responding to a visual stimulus (Dinges, et al., 1985). It typically lasts for 10-minutes, but can also be effective in measuring the components of sleep after as little as 1 minute (Gartenberg & Parasuraman, 2011).

Recently, Shaw, Warm, Finomore, Tripp, Matthews, Ernest, & Parasuraman (2009) used transcrannial doppler (TCD) to determine that the declining performance characteristic of vigilance tasks was due to a lack of resources and not from changes in systemic activity. Lim et al. (2010) found that when participants were sleep deprived, a 20-minute time-on-task could detect the vigilance decrement using the PVT, yet the task was sensitive to sleep deprivation in as little as 2 minutes. This suggests that the PVT is resource demanding, but that its sensitivity to detecting fatigue does not rely on the vigilance decrement function. In support of this, tasks that do not involve the decrement, such as brief cognitive tasks that require speed of cognitive throughput, working memory, and other aspects of attention have been found sensitive to sleep deprivation (Mallis, et al., 2008).

Based on the resource theory of vigilance, tasks that require more resources, such as successive tasks that tax working memory, will be more sensitive to detecting the vigilance decrement than the traditional PVT (Shaw, Warm, Finomore, Tripp, Matthews, Ernest, & Parasuraman, 2009). Perhaps they will also be more sensitive to detecting grogginess. Yet, when developing a neuroergonomic task that requires more resources, it is important to consider possible learning effects because learning can attenuate the task's sensitivity to detecting grogginess.

We developed and tested a simultaneous vigilance task, a successive vigilance task, and two N-back tasks on an iPhone platform and a Desktop platform in order to investigate potential issues involved in administering the Mind Metrics application in naturalistic environments.

3   MIND METRICS SMARTPHONE APP

We developed an neuroergonomic smartphone application that detects alertness level and tested the application on both an iPhone (with 320px by 480 px dimensions) and a desktop computer. The application included three types of tasks: vigilance tasks, memory tasks, and combined vigilance and memory tasks. After completing a task the participant received real-time feedback on their performance. User performance was saved to a viewable table where it could be exported.

3.1   Vigilance, Working Memory, and Combined Tasks

The PVT is a simultaneous vigilance task that requires participants to respond when a sun appears in the center of the screen. There is a stimulus onset window of 10,000 milliseconds, for which the sun randomly appears for 1,000 milliseconds. A total of 60 trials were run in the 10-minute version of this task.

The 10-minute spatial discrimination vigilance task (SDVT) is a successive vigilance task that involved discerning the distance between two stimuli. The stimuli consisted of a stationary cloud positioned in the center of the screen and a moon that appeared at one of two distances from the cloud (either 110 pixels or 130 pixels). The moon was presented close to the cloud 80% of the time (noise) and far from the cloud 20% of the time (signal). A response was only required when the moon was far from the cloud. Each trial lasted 4,300 milliseconds. The cloud remained present for the entire duration of each trial. The trials began with 1,800 milliseconds of inter-trial interval where only the cloud was presented. This was followed by the presentation of the moon stimulus, which lasted for 300 milliseconds. After the presentation of the stimulus the participant was given 1,900 milliseconds to respond. Feedback was then presented on whether the answer was correct or incorrect. The feedback lasted for 200 milliseconds.

The 10-minute color n-back (CNB) and spatial n-back (SNB) were 2-back working memory tasks. Both tasks required participants to determine if a set of lightning bolts was the same or different from a set of lightning bolts that occurred two trials previously. Each trial lasted for 4,300 milliseconds. Each trial began with 2,100 milliseconds interval where the cloud and the lightning bolts were present and the participant had the opportunity to respond. The feedback was then presented where the answer was either or incorrect. This feedback lasted for 200 milliseconds. There was an inter-trial interval of 1900 milliseconds where only the clouds were presented. The participant either indicated that the current stimulus was the same as the lightning pair that appeared two trials before or different from the lightning pair that appeared two trials before.

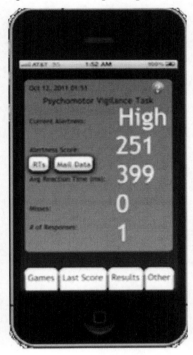

The CNB and SNB differed based on the modality of the stimuli. The CNB involved holding a color representation in working memory, while the SNB required holding a spatial representation in working memory. In the CNB two lightning bolts were presented that were different colors. In the SNB two lightning bolts were presented that were oriented in different spatial locations.

The combined vigilance and working memory task had an identical stimulus onset as the CNB and SNB, with the exception that the SDVT was administered in conjunction with the working memory tasks. This task was designed to be a task with a high task load.

Figure 1. Mind Metrics feedback screen of the PVT.

3.2 Feedback

Figure 1 illustrates the feedback that participants receive after completion of the task. Participants get information on the number of trials completed, their average accuracy, reaction time, and a score that combines accuracy and reaction time. Based on other scores that the participant received on the task, feedback is provided on alertness. This provides users with a real time measure of their alertness.

4 METHOD

4.1 Participants

48 George Mason University students voluntarily participated for course credit. Each participant had normal or corrected vision. The sample consisted of 26 men and 22 women. The average age of participants was 20.25 years with a standard deviation of 3.41 years. Two participant's data were eliminated in the iPhone condition due to loss of Internet connection during the experiment.

4.2 Design and Procedure

The experiment was a mixed design with device between groups and task within groups. The experiment was conducted using a desktop computer running E-Prime software and a 3rd generation iPod touch using an application running the iOS 4.0 SDK. The desktop and the iPod touch were programmed with four tasks where all participants experienced each task: a Psychomotor Vigilance Task (PVT), a Spatial Discrimination Task (SDVT), a Color N-Back (CNB), and a Spatial N-Back (SNB). Participants were assigned to platform conditions using Latin-squared randomized. The desktop-platform tasks and the iPod-platform tasks were identical, with the exception that in the desktop-based task participants responded with a button-press.

Each task began with instructions. The instructions were followed by a 1-minute practice session. The practice continued until criteria were met, where the criteria for each task differed due to different likelihoods of responding correctly. The PVT has very little practice effect the PVT due to the simplicity of the task, making it not necessary to train to criteria for this task. For the SDVT, if the participant never gave a response, they were able to obtain a score of 80%. Therefore, the criterion for the SDVT was set to 90%. For the N-Back tasks participants had a 50% chance of responding correctly. The criteria for the N-Back tasks were thus set to 80% correct.

After criterion was met, participants were reminded to respond as quickly and accurately as possible and to respond using their index finger with the finger hovering above the response interface. The 10-minute task then began.

Upon completion of each task, the perceived workload of the task was measured by administering the NASA-TLX (Hart & Staveland, 1988). Once the NASA-TLX was finished, the participant was given a short 5-minute break.

4.3 Measures

Reaction time and accuracy were used as measured. As is customary with analysis of vigilance tasks, each 10-minute task was divided into 5 blocks of 2-minutes each. This enabled for the detection of changes in performance as the task progressed.

5   RESULTS AND DISCUSSION

The vigilance decrement was calculated by measuring reaction time in 2-minute block intervals, as the task progressed. For the PVT, this meant that there were 12 trials per interval. For the SDVT, CNB, and SNB there were 24 trials per interval.

For the PVT, a mixed ANOVA was run with block as a within groups factor and device type as a between groups factor. There was a main effect of block on reaction time, where participants performed worse as the task progressed, $F(4, 172) = 9.93$, $p < .05$. There was a main effect of device where participants were slower on the PVT when using an iPhone ($M = 489.76$ ms, $SD = 79.29$ ms) than when using a desktop ($M = 348.67$, $SD = 46.75$), $F(1, 43) = 54.49$, $p < .05$ (see Figure 2). There was no interaction between block and device condition, $F(4, 172) = 9.93$, $p < .05$. The finding that the overall reaction time was slower for the iPhone condition than the desktop condition suggests that the iPhone platform samples reaction times at a slower rate than a desktop computer running E-Prime software. However, the main effect of interval time and the lack of an interaction suggested that the simultaneous vigilance tasks were sensitive to the vigilance decrement despite these differences in the sample rates between devices.

For the SDVT, there was only reaction time data collected for the desktop condition. A simple within groups ANOVA was conducted on the 2-minute block intervals. As expected, over time participants performed worse on the task, $F(4, 84) = 2.73$, $p < .05$ (see Figure 2). This suggested that, successive vigilance tasks were sensitive to the vigilance decrement after a 10 minute task period.

It was important to determine if the two types of n-back tasks were equated on difficulty because these tasks are used for the combined vigilance and working memory task. Reaction time data was collected for the desktop condition and accuracy data was collected for both the desktop and iPhone condition. There was no difference in reaction time between the CNB ($M = 489.76$ ms, $SD = 79.29$ ms) and SNB ($M = 489.76$ ms, $SD = 79.29$ ms), $t(22) = 0.62$, $p = .54$. There was no difference in accuracy between the CNB ($M = 83.41\%$, $SD = 11.53\%$) and SNB ($M = 83.02\%$, $SD = 10.04\%$), $t(41) = 0.34$, $p = .74$. Since these tasks were equated for difficulty, this suggests that any differences in performance for the combined vigilance and working memory task should be related to the modalities of the n-backs (i.e. spatial vs color).

A simple within groups ANOVA was conducted on block. For the CNB there was no difference in how participants performed over time, $F(4, 88) = 0.70$, $p = .59$ (see Figure 2), but for the SNB participants responded more quickly over time, $F(4, 88) = 3.67$, $p < .05$ (see Figure 2). This suggested that overall difficult was similar between the two N-Back tasks. While there was a general trend of faster response times, this effect was only found for the SNB.

A global measure of perceived workload was determined using the NASA-TLX and a mixed ANOVA with task as the within groups factor and device as the between groups factor. There was no difference in perceived workload between the iPhone conditions and the Desktop conditions, $F(1, 45) = 0.79$, $p = .38$. Perceived workload differed based on task condition, $F(1, 45) = 70.41$, $p < .05$. Post-hoc tests were conducted using the Benjamini Hochberg correction method. All groups were significantly different from all other groups ($p < .05$), with the exception of the SNB and the CNB ($p = .33$). Participants rated the CNB ($M = 66.94$, $SD = 16.23$) and SNB ($M = 65.21$, $SD = 15.85$) as more difficult than the SDVT ($M = 57.70$, $SD = 19.70$) and the PVT ($M = 42.82$, $SD = 24.55$). There was no interaction between task condition and device, $F(1, 45) = 1.94$, $p = .17$. This suggested that the participant perceived the n-backs to be the most difficult, followed by the SDVT, and the PVT.

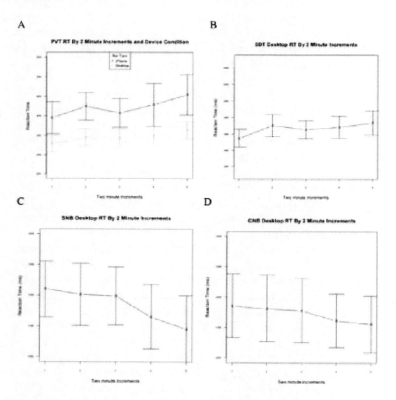

Figure 2. Graph of RT broken into 2 minute blocks. A = PVT, B = SDVT, C = SNB, D = CNB.

GENERAL DISCUSSION

We developed a neuroergonomic smartphone application called Mind Metrics, which provides people with real-time measures of their arousal state. The application includes a simultaneous vigilance task, successive vigilance task, working memory tasks, and combined vigilance and working memory tasks. Users of the application can set the duration and difficulty of the tasks, get feedback based on their own unique individual performance, and save and export their data.

In this experiment, the PVT, SDVT, CNB, and SNB were tested on both desktop and iPhone devices. Both devices detected the vigilance decrement, but the iPhone registered slower reaction times than the desktop. Possible explanations for this include slowing of response due to the touch interface and the iPhone software more slowly registering touch presses than a desktop computer. Nonetheless, the iPhone was used to measure the vigilance decrement, which will help foster naturalistic data collection, a requirement of neuroergonomics.

Another requirement of neuroergonomics measurement tools are non-invasiveness to a worker's daily routine. A vigilance task that incurs more resource demands on the user improves the time sensitivity of these tasks to measure the vigilance decrement (Warm, & Dember, 1998). As a result, increasing resource demands may be more sensitive to changes in an individual's arousal system.

We discovered that memory tasks such as the N-back, have higher perceived resource demands than the vigilance tasks administered in this study. However, no decrement in performance was found in the n-back tasks. The reason for this may be that these tasks do not have the characteristics of a vigilance task and that learning plays a larger role in these tasks. Administering tasks repeatedly could reduce these learning effects.

The neuroergonomic smartphone application developed in this paper was still in its testing phase. A limitation of this study was that the tasks were not administered repeatedly. In future research, the application will be applied to naturalistic environments and administered repeatedly. This will enable for real-time detection of the worker's arousal system, which can be used to prevent accidents and the negative consequences of accidents.

REFERENCES

Baehr, E., Revelle, W., & Eastman, C. (2000). Individual differences in the phase and amplitude of the human circadian temperature rhythm: with an emphasis on morningness-eveningness. *Journal of Sleep Research, 9*, 117-127.

Balkin, T., Rupp, T., Picchioni, D., & Wesensten, N. (2008). Sleep Loss and Sleepiness. *Chest, 134*, 653-660.

Boksem, M. A., Meijman, T. F., & Lorist, M.M. (2005). Effects of mental fatigue on attention: an ERP study. *Brain Res. Cogn. Brain Res. 25*, 107-116.

Bonnet, M. H. (1991). The Effect of Varying Prophylactic Naps on Performance, Alertness and Mood throughout a 52-Hour Continuous Operation. *Sleep, 14*, 4, 307-315.

Caggiano, D., & Parasuraman, R. (2004). The role of memory representation in the vigilance decrement. *Psychonomic Bulletin and Review, 11*, 5, 932-937.

Davies, D. & Parasuraman, R. (1982).*The Psychology of Vigilance.* London:Academic Press.

Dinges, D. F., Orne, M. T., Whitehouse, W. G., & Orne E. C. (1987). Temporal placement of a nap for alertness: contributions of circadian phase and prior wakefulness. *Sleep, 10*, 313–329.

Dinges, D. F., & Powell, J.W. (1985). Microcomputer analyses of performance on a portable, simple visual RT task during sustained operations. *Behav Res Meth Instr Comp, 17*, 652–655.

Gartenberg, D. & Parasuraman, R. (2010). Understanding Brain Arousal and Sleep Quality Using a Neuroergonomic Smart Phone Application. In Marek, T., Karwowski, W., &

Rice, V. (Eds.), Advances in Understanding Human Performance, *3rd International Conference on Applied Human Factors and Ergonomics* (pp. 210-220).

Hart, S. G., & Staveland, L. E. (1988). Development of NASA-TLX (task load index): Results of empirical and theoretical research. In P. A. Hancock & N. Meshkati (Eds.), Human mental workload. (pp. 139-183). Oxford, UK: North-Holland.

Helton, W., Dember, W., Warm, J., & Matthews, G. (2000). Optimism, pessimism, and false failure feedback: Effects on vigilance performance. *Current Psychology, 18*, 311-325.

Lim, J., Wi, W., Wang, J., Detre, J. A., Dinges, D. F., & Rao, H. (2010). Imaging brain fatigue from sustained mental workload: An ASL perfusion study of the time-on-task effect. *Neuroimaging, 49*, 3426-3435.

Nuechterlein, K., Parasuraman, R., & Jiang, Q. (1983). Visual sustained attention: Image degradation produces rapid sensitivity decrement over time. *Science, 220*, 327-329.

Mackworth, J. F. (1968). Vigilance, arousal, and habituation. *Psychol. Rev., 75*, 308-322.

Mallis, M., Banks, S., & Dinges, D. Sleep and circadian control of neurobehavioral functions. Ed. Parasuraman, R., & Rizzo, M. *Neuroergonomics: The Brain at Work.* New York: Oxford Univsersity Press, 2007.

Matthews, G., Davies, D. Westerman, S. J., & Stammers, R. B. (2000). *Human performance: cognition, stress, and individual differences*. Philadelphia: Taylor and Francis.

Maquet, P. (2001). The role of sleep in learning and memory. *Science, 294*, 1048–1052.

Mori, C., Bootzin, R., Buysse, D., Edinger, J., Espie, C., & Lichstein, K. (2006). Psychological and Behavioral Treatment of Insomnia: Update of the Recent Evidence (1998-2004). *Sleep, 29,* 11.

Parasuraman, R. (1985). Sustained attention: A multifactorial approach. In M. I. Posner & O. S. Marin (Eds.) *Attention and Performance XI.* (pp. 493-511). Hillsdale, New Jersey: Erlbaum Associates.

Parasuraman, R. (1986). Vigilance, monitoring, and search. In K. Boff, L. Kaufman, & J. Thomas (Eds.), *Handbook of perception and human performance. Vol. 2: Cognitive processes and performance* (pp. 43.1-43.39). New York: Wiley.

Parasuraman, R., (2003). Neuroergonomics: research and practice. *Theor. Issues. Ergon. Sci. 4*, 5-20.

Rizzo, M., Robinson, S., & Neale, V. The Brain in the Wild: Tracking Human Behavior in Naturalistic Settings. Ed. Parasuraman, R., & Rizzo, M. *Neuroergonomics: The Brain at Work.* New York: Oxford University Press, 2007.

Shaw, T. H., Warm, J. S., Finomore, V., Tripp, L., Matthews, G., Ernest, W., & Parasuraman, R. (2009). Effects of sensory modality on cerebral blood flow velocity during vigilance. *Neuroscience Letters, 461*, 207-211.

Stutts, J. C., Wilkins, J. W., Osberg, S. J., & Vaughn, B. V. (2003). Driver risk factors for sleep-related crashes. *Accident Analysis and Prevention, 35,* 321-331.

Van Dongen, H. P., Maislin, G., Mullington, J. M., & Dinges, D. F. (2003). The cumulative cost of additional wakefulness: Dose-response effects on neurobehavioral functions and sleep physiology from chronic sleep restriction and total sleep deprivation. *Sleep, 26,* 117-126.

Warm, J.S., Parasuraman, R., & Mathews, G. (2008). Vigilance requires hard mental work and is stressful. *Hum. Factors, 50*, 433-441.

Warm, J. S., & Dember, W. N. (1998). Tests of vigilance taxonomy. In R. R. Hoffman, M. R. Sherrick, & J. S. Warm (Eds.), Viewing psychology as a whole (pp. 87-112). Washington, DC: American Psychological Association.

CHAPTER 26

fNIRS and EEG Study in Mental Stress Arising from Time Pressure

Shyh-Yueh CHENG 1, Chao-Chen LO 2, Jia-Jin CHEN 2

1 Chia-Nan University of Pharmacy and Science, Tainan, 717, Taiwan, ROC
2 National Cheng Kung University, Tainan, 701, Taiwan, ROC
csy196166@yahoo.com.tw

ABSTRACT

In our competitive society, efficiency is of central concern. The pace of life is gradually becoming faster. Whether official business at the workplace or private business at home, we may face stress and be bothered with anxiety and emotionality. The result may cause physical and mental damage. The stress employees encounter at work may essentially endanger their health; moreover, it can induce human error that may cause an accident. Past research suggests 70% - 90% of system failures result directly or indirectly from human error due to work stress. The result of stress may thus severely impair the system and the worker. The aim of this study was to investigate the relationship between different levels of time pressure and brain activities measured by fNIRS (functional near-infrared spectroscopy) and electroencephalogram (EEG). Ten university students participated as volunteer subjects in this study. The fNIRS results showed that brain activity significantly decreased with increasing time pressure in the frontal and left occipital areas; with increased time pressure, the control of blood vessels within the brain decreased. The EEG basic indices θ and α at all recording sites presented to increased and decreased, respectively, with the level of time pressure. The index β tended to increase from time pressure level 1 to level 2, while decreasing from time pressure level 2 to level 3. This demonstrated that subjects failed to relax, got more tired, and appeared less attentive when facing high levels of time pressure. The amplitude of the ERP (event related potential) registered from the occipital lobe of participants decreased with time pressure. The results suggest that information processing ability is decreased under time pressure. NASA-Task Load Index (TLX) rating scales in three mental arithmetic tasks showed no significant differences, which suggests that subjective assessment via questionnaires is not sensitive for measurement of mental stress and reveals the valuable application of using physiological measurement.

Keywords: mental stress, functional Near Infrared Spectroscopy (fNIRS), electroencephalogram (EEG), event relate potential (ERP)

1 INTRODUCTION

In our competitive society, efficiency is of central concern. The pace of life is gradually becoming faster. Whether official business at the workplace or private business at home, stress is becoming a part of our lives. Stress, especially mental stress, may result in anxiety and emotionality. Mental stress from the workplace may cause physical and mental harm, even increase the probability of accidents. Lin and Hwang (1992) studied workers' physical and mental stressors and found they usually arise from task demands at the workplace. As the performance of workers becomes mismatched with task demands, the workplace brings stress to workers. The stress employees encounter at work may essentially endanger their health; moreover, it can induce human error that may cause an accident. Past research suggests that 70% - 90% of system failures result directly or indirectly from human error due to work stress (Lin & Hwang, 1992).

Thus, it is important to develop a system to monitor mental stress to prevent potential accidents. Among currently available functional neural imaging approaches, Electroencephalography and fNIRS are non-invasive, portable methods to evaluate mental processing. The advantage of fNIRS and EEG is their high temporal resolution. The temporal resolution of EEG can be as high as milliseconds, whereas that of fNIRS can be within seconds. Also these two modalities are comparably inexpensive and portable. NIRS is an optic method to measure hemodynamic in the tissue. By using different wavelengths, oxy-hemoglobin and deoxy-hemoglobin can be measured through calculating the relationship between absorption coefficient, scattering coefficient, and light attenuation. In recent years, various cognitive tasks (Nishimura, et al., 2009; Kubo, et al., 2008; Azechi, et al., 2009) were performed to measure the prefrontal cortex to understand the relationship between hemodynamic response of the prefrontal cortex and different events. Studies of visual tasks were also performed to understand visual cortex hemodynamic changes. By placing a visual stimulus in front of subjects, changes in oxyhemoglobin concentration in the occipital lobe can be induced (Meek, et al., 1995). Moreover, if subjects were asked to pay attention to the stimulus, activity of the visual cortex would be enhanced (Kojima & T. Suzuki, 2010).

Electroencephalogram contains spontaneous brain activity and ERPs. Brain oscillations are the interaction of neuronal networks within the brain, including various frequencies (Fisch, 1991). Delta (δ) band is the frequency range up to 4 Hz and it happens in a deep sleep state. Theta (θ) band presents the spectral power value from 4 to 8 Hz and alpha (α) band is the power spectrum from 8 to 13 Hz. Beta (β) is the frequency range from 13 to 20 Hz. Study has shown that with increasing workload, θ activity increases in the frontal lobe and α activity decreases in the occipital lobe (Holm, et al., 2009). The ERP is a transient series of voltage oscillations in the brain recorded from scalp EEG following a discrete event. This ERP component has been found to reflect the further processing of relevant information (i.e., stimuli that require a response) (Lange et al., 1998; Okita et al., 1985; Wijers et al, 1989a, b). In stimulus-locked ERP, the P300 has been defined as the most positive peak in a window between 200 and 500 milliseconds (Ullsperger et al., 1986, 1988). The P300 component is useful to identify the depth of cognitive information processing. It has been reported that the P300 amplitude elicited by mental task loading decreases with increases in the perceptual/cognitive difficulty of a task (Donchin, 1979; Isreal et al., 1980a, b; Kramer et al., 1983, 1985; Mangun and Hillyard, 1987; Ullsperger et al., 1986, 1988). On the other hand, P300 has been shown to decrease in amplitude and increase in latency.

Fewer studies investigate the relationship between different workloads and brain activities from hemodynamic and electricity of the brain. Stroop task was used to demonstrate these two modalities simultaneously in prefrontal cortex (Zhai, et al., 2009). Another study indicated the relationship between electrophysiology and hemodynamic by giving an electric stimulus on the median nerve of right hand (Takeuchi, et al., 2009). Nevertheless, there is little information about mental stress and brain activities measured by the combination of fNIRS and EEG. So the aim of this study is to investigate the relationship between different levels of time pressure and brain activities measured by fNIRS and EEG.

2 METHODS AND MATERIALS

2.1 Subject

Ten university students including 9 males and 1 female with age ranging from 18 to 24 participated as volunteer subjects. They had normal hearing and normal or corrected-to-normal vision. Each participant met all the inclusion criteria: no medical concerns, psychiatric issues, or head injuries, and they could not be using any medications or drugs. An informed written consent form was obtained from all participants after the study procedure was explained and the laboratory facilities were introduced to them. They were paid for their participation in the study.

2.2 Experiment protocol

The subjects were required to record their fNIRS and EEG before starting the experimental session. Following the fNIRS and EEG measurements in the rest condition for 5 min, self-report assessments of task loading were obtained using the NASA-TLX rating scale (Hart & Staveland, 1988). To acquire ERP, the subject performed a modified Eriksen flanker task with word stimuli (Eriksen & Eriksen, 1974) replaced by arrow stimuli for 5 min. The stimuli were presented on a computer screen (15 inches) with a dark background and with a viewing distance of 80 cm. After the measurement of the ERPs had been finished, the subject conducted an experiment task. There were three levels of time pressure for the experiment tasks, with 9, 7, and 5 seconds to respond to the mental arithmetic as three levels of time pressure, as task 1, 2, and 3 individually. The levels of time pressure varied randomly. In the experiment tasks, every subject was asked to mentally add two three-digit numbers for three sessions, each 15 minutes; numbers were presented on a black screen and subjects had to type their answers. Similar EEG recordings were conducted immediately after the completion of each level of the experiment task.

2.3 fNIRS data collection

NIRS measurement was conducted with a frequency domain NIRS system (ISS Imagent). The light detectors were placed at F3, F4, O1, and O2 electrodes in the international 10–20 montage (Andreassi, 2000), with light source set around them with three-centimeter-distance. Two different wavelengths (830nm and 690nm) were used in order to get oxy-hemoglobin and deoxy-hemoglobin. There are overall 14 channels. When the visual stimulus appeared, this would send a marker to the NIRS computer. According to these markers, six different trials of calculations can be averaged to gain a general trend of oxy-hemoglobin. Homer was used to analyze the NIRS data. Mean oxy-hemoglobin changes were calculated to quantify the activation of the brain.

2.4 EEG data collection

During the task performance, EEG was recorded using an electrode cap (Quick-Cap, Compumedics NeuroScan, El Paso, Tex), with Ag/AgCl electrodes placed at the same electrode sites the NIRS measuring used, with an electronically linked mastoids reference. Two Ag/AgCl electrodes, 2 cm above and 2 cm below the left eye, recorded vertical electrooculogram (EOG), and 2 electrodes 1 cm external to the outer canthus of each eye recorded horizontal EOG. A ground electrode was placed on the forehead. Electrode impedances were kept below 10 kΩ. The EEG and EOG were amplified by SYNAMPS amplifiers (Neuroscan, Inc.) and sampled at 500 Hz. The EEG epochs were then corrected for eye movement by using the Ocular Artifact Reduction (Semlitsch et al., 1986) command of SCAN 4.3 (Neuroscan, Inc.) and then underwent movement-artifact detection by using the Artifact Rejection command.

2.1 Event related potential

The latency of each ERP component was defined as the time between the onset of the arrow array from modified Eriksen flanker task and the time when the peak value appeared for stimulus-locked ERP. The amplitude and latency measures for P300 were derived from the stimulus-locked ERP recorded at F3, F4, O1, and O2 electrodes, respectively. It is noted that the EEG epochs of the trials with omitted responses or with reaction times longer or shorter than twice the value of the standard deviation for reaction time were not included in the stimulus-locked ERP.

3 RESULTS

3.1 NIRS measurement

A typical hymodynamic response during the experiment task showed an increase in oxyhemoglobin and decrease in deoxyhemoglobin, as shown in Fig. 1. The region between two vertical lines is the time of calculation and response. Cortical activation of different brain regions was extracted from hymodynamic responses and plotted in the order of the levels of time pressure. Fig. 2 shows that the greatest oxyhemoglobin changes occurred in task 3. The pattern of cortical activity level is similar in frontal (F3, F4) and left occipital (O1) areas. All these three sites showed a decrease in task 2 compared with task 1 and in task 2 compare with task 3. A significant difference of cortical activation between task 3 and task 1 is depicted in Fig. 2. It showed that with increasing time pressure, the control of the blood vessels of the brain decreased.

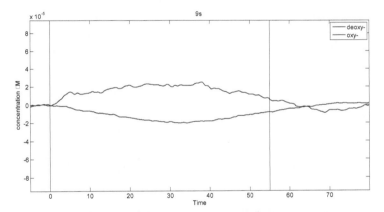

Fig. 1 A typical example of hemodynamic response during mental processing. The red line is oxyhemoglobin and the blue line is deoxyhemoglobin. The region between two vertical lines is the time of calculation and response.

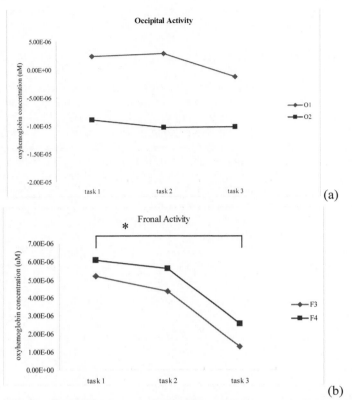

Fig. 2 Cortical activation of different levels of time pressure in (a) frontal region and (b) occipital region. There was a significant difference between task 3 and task 1. (*: P<0.05)

3.2   EEG measurement

The EEG basic indices θ and α at all recording sites (especially O2) present an increase and decrease, respectively, as depicted in Fig. 3 (a) and (b), with the level of time pressure. The index β tends to increase from time pressure level 1 to level 2, while decreasing from time pressure level 2 to level 3. The results demonstrate that subjects failed to relax, got more tired, and appeared less attentive when faced with a high level of time pressure.

3.3   ERP analysis

The P300 amplitude at all recording sites tended to increase from time pressure level 1 to level 2, while significantly decreasing from time pressure level 2 to level 3 (especially at O1). A similar tendency was found in P300 amplitude. It revealed that as subjects undertook tasks under time pressure, this in turn decreased the cognitive response. This phenomenon revealed a decreased depth of cognitive information processing and a decreased level of attention.

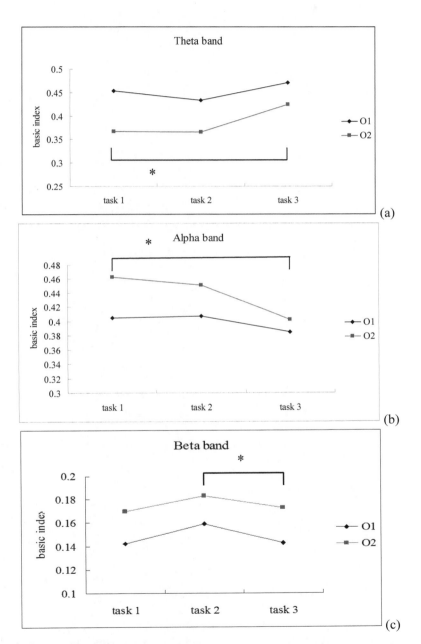

Fig. 3 Comparison of EEG basic indices of beta (a), alpha (b), and theta (c) in occipital lobe under different time pressure levels. (*: P<0.05)

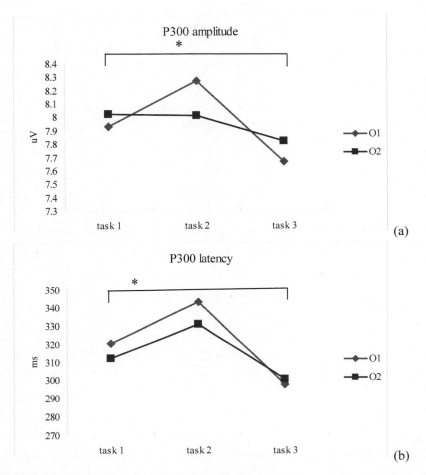

Fig. 4 Comparison of P300 amplitude (a) and latency (b) in occipital region under different time pressure levels. (*: P<0.05)

3.4 NASA-TLX rating scale

NASA-TLX rating scales in the 3 mental arithmetic tasks showed no significant differences. These results suggest that subjective assessment via questionnaires is not sensitive in detecting mental pressure.

4 DISCUSSION AND CONCLUSION

This study examined a combination of fNIRS and EEG recordings for assessing the mental stress arising from time pressure. In NIRS findings, the results showed that with an increase in time pressure, the activity of the brain decreased; thus, with increasing time pressure, the control of blood vessels within the brain decreased.

The results imply that activities of the brain were significantly decreased with increasing time pressure. On the other hand, EEG recordings found that subjects failed to relax, got more tired, and appeared less attentive when faced with high level of time pressure. This indicates that if time pressure is too high, cortical activation decreases. Both measurements demonstrated a good correlation in the results. The study results show that the EEG basic index of α band in task1 was higher than task3. This demonstrated that time pressure increased with response speed of mental arithmetic. The EEG basic index of θ band in task3 was higher than task1. This suggests that a rapid response induced stress on subjects and a tardy response was easy for subjects. The EEG basic index of β band, i.e. alert level, increased with response speed of mental arithmetic. This suggests that subjects needed to pay more attention to finish the mental arithmetic under time pressure, but as they faced a high level of time pressure their attention would lower. The P300 amplitude at all recording sites tended to increase from time pressure level 1 to level 2, while significantly decreasing from time pressure level 2 to level 3 (especially at O1). These results suggest that information processing ability is decreased under a high level of time pressure. NASA-TLX rating scales in the 3 mental arithmetic tasks showed no significant differences. This suggests that subjective assessment via questionnaires is not sensitive for measurement of mental stress and reveals the valuable application of using physiological measurement.

ACKNOWLEDGMENTS

The authors would like to thank the National Science Council (NSC) of Taiwan for financially supporting this research under contract NSC 99-2221-E-041-001-.

REFERENCES

Andreassi, J.L., 2000. *Psychophysiology: Human Behavior and Physiological Response* (4th ed.), New Jersey: Lawrence Erlbraum Associates.
Azechi, M., et al., Nov 4, 2009. Discriminant analysis in schizophrenia and healthy subjects using prefrontal activation during frontal lobe tasks: A near-infrared spectroscopy, *Schizophr Res*.
Donchin, E., 1979. Event-related brain potentials: a tool in the study of human information processing. In: Begleite, H. (Ed.), *Evoked Brain Potentials and Behavior*. Plenum Press, New York.
Eriksen, A. and C. W. Eriksen, 1974. Effects of noise letters upon the identification of a target letter in a nonsearch task, *Percept. Psychophys.*, 16: 143-149.
Hart, S. G. and L. E. Staveland, 2009. Development of NASA-TLX (Task Load Index): results of experimental and theoretical research, in: P. A. Hancock and N. Meshkati (Eds.), *Human Mental Workload*, Amsterdam: Elsevier, 39-183.
Holm, A., et al., 2009. Estimating brain load from the EEG, *Scientific World Journal*, vol. 9, pp. 639-51.
Isreal, J. B., G. L. Chesney, C. D. Wickens, E. Donchin, 1980. P300 and tracking difficulty: evidence for multiple resources in dual-task performance, *Psychology*, vol. 17, pp. 259-273.

Isreal, J. B., C. D. Wickens, G. L. Chesney, E. Donchin, 1980. The event-related brain potential as an index of display monitoring workload, *Human Factors*, vol. 22, pp. 211-224.

J. Fisch, 1991. *Sphelmann's EEG Primer* (2nd ed.), Amsterdam: Elsevier Science BV.

Kojima, H. and T. Suzuki, 2010. Hemodynamic change in occipital lobe during visual search: visual attention allocation measured with NIRS, *Neuropsychology*, vol. 48, pp. 349-52.

Kramer, A. F., C. D. Wickens, E. Donchin, 1983. An analysis of the processing demands of a complex perceptual-motor task, *Human Factors*, vol. 25, pp. 597-622.

Kramer, A. F., C. D.Wickens, E. Donchin, 1985. Processing of stimulus properties: evidence for dual-task integrality, *Journal of Experimental Psychology, Human Perception and Performance*, vol. 11, pp. 393-408.

Kubo, M., et al., 2008. Increase in prefrontal cortex blood flow during the computer version trail making test, *Neuropsychobiology*, vol. 58, pp. 200-10.

Lange, J. J., A. A. Wijers, L. J. Mulder, G. Mulder, 1998. Color selection and location selection in ERPs: differences, similarities and Fneural specificity, *Biol. Psychol.*, vol. 48 (2), pp. 153-182.

Lin, Y. L., and Hwang, S. L., 1992. The application of the loglinear model to quantify human error. *Reliability Engineering and System Safety*, 37(2): 157-165.

Mangun, G. R., S. A. Hillyard, 1987. The spatial allocation of visual attention as indexed by event-related brain potentials, *Human Factors*, vol. 29, pp. 195-211.

Meek, J. H., et al., 1995. Regional Changes in Cerebral Haemodynamics as a Result of a Visual Stimulus Measured by near Infrared Spectroscopy, *Proceedings* of the Royal Society of London Series B-Biological Sciences, vol. 261, pp. 351-356.

Nishimura, Y., et al., 2009. An event-related fNIRS investigation of Japanese word order, *Exp. Brain Res.*

Okita, T., A. A. Wijers, G. Mulder, L. J. M. Mulder, 1985. Memory search and visual spatial attention: an event-related brain potential analysis, *Acta Psychol.*, vol. 60, pp. 263-292.

Semlitsch, H. V., Anderer, Schuster, P., P. and Presslich, O., 1986. A solution for reliable and valid reduction of ocular artifacts, applied to the P300 ERP, *Psychophysiology*, 23: 695-703.

Ullsperger, P., A.-M. Metz, H. G. Gille, 1988. The P300 component of the event-related brain potential and mental effort, *Ergonomics*, vol. 31, pp. 1127-1137.

Ullsperger, P., U. Neuman, H.-G. Gille, M. Pietschann, 1986. P300 component of the ERP as an index of processing difficulty. In: Flix, F., Hagendor, H. (Eds.), *Human Memory and Cognitive Capabilities*. North-Holland, Amsterdam.

Wijers, A. A., W. Lamain, S. Slopsema, G. Mulder, L. J. M. Mulder, 1989. An electrophysiological investigation of the spatial distribution of attention to coloured stimuli in focussed and divided attention conditions, *Biol. Psychol.*, vol. 29, pp. 213-245.

Wijers, A.A., G. Mulder, T. Okita, L. J. M. Mulder, M. K. Scheffers, 1989. Attention to colour: an ERP-analysis of selection, controlled search, and motor activation, *Psychophysiology*, vol. 26 (1), pp. 89-109.

Zhai, J., et al., Nov, 2009. Hemodynamic and electrophysiological signals of conflict processing in the Chinese-character Stroop task: a simultaneous near-infrared spectroscopy and event-related potential study, *J Biomed Opt*, vol. 14, p. 054022.

CHAPTER 27

Availability and Future Prospects of Functional Near-infrared Spectroscopy (fNIRS) in Usability Evaluation

Hiroaki Iwasaki*, Hiroshi Hagiwara**

*Graduate School of Science and Engineering
**College of Information Science and Engineering
Ritsumeikan University
Siga, Japan
ci001076@ed.ritsumei.ac.jp

ABSTRACT

In recent years, usability evaluations have evolved into comprehensive assessments that combine objective metrics and subjective evaluations. It has been shown that the prefrontal cortex is responsible for higher brain functions such as cognition, judgment, and attention. Activation of the prefrontal cortex during use of a product can therefore be regarded as an indicator of the amount of executive function required for product use, and is considered a measure of usability. In the present study, we used functional near-infrared spectroscopy (fNIRS) to measure hemodynamic changes in the prefrontal cortex during performance of repetitive mousing tasks on a computer. The usability of these tasks was varied by adjusting the difficulty of the task. We evaluated the effects of usability on brain activity on oxygenated hemoglobin (oxyHb) levels in the prefrontal cortex and dorsolateral area 46. Eleven participants (9 male, 2 female) completed a click mousing task and a drag mousing task on two separate days. We quantified the overall change in oxyHb (δ_{oxyHb}) levels during task performance. Positive δ_{oxyHb} values indicate an overall increase in oxyHb levels across the duration of the task, whereas negative

δ_{oxyHb} values indicate an overall decrease in oxyHb levels. We found negative δ_{oxyHb} values in drag mousing tasks with the largest target size, indicating that oxyHb levels decreased during the best-quality usability task. We found positive δ_{oxyHb} values in other usability tasks. Additionally, δ_{oxyHb} values tended to increase when the usability deteriorated. These results suggest that fNIRS may be useful in usability evaluations.

Keywords: usability, fNIRS, prefrontal cortex, oxyHb, brain activation

1 INTRODUCTION

In recent years, the development of multifunctional electronic and information devices has complicated the use of the equipment. Complex product specifications lead to excessive demands on the user and can lead to a sense of discomfort and frustration. Manufacturers are beginning to consider the quality of usability, and are trying to differentiate the usability of equipment they produce.

Traditionally, usability has been evaluated by subjective evaluation and behavioral skills. However, these assessments can be confounded by the user's experience and knowledge, and evaluation results are not always valid. More recently, usability evaluation has comprised a comprehensive assessment of a combination of subjective and objective indicators.

Gaining a picture of human brain activity is becoming possible with the development of technologies such as functional magnetic resonance imaging (fMRI) and positron emission tomography (PET) that allow non-invasive assessment of brain activity. However, the use of fMRI and PET technologies in usability assessments is limited as these devices are difficult to use when the subject is in a standing or sitting position, forcing the subject to be lying down during the assessment. Functional near-infrared spectroscopy (fNIRS) has attracted attention because it is possible to measure brain activity in a more natural state by using near-infrared light. We believe that fNIRS has potential to be used in usability assessments. This study was performed to assess the use of fNIRS to evaluate usability.

2 EXPERIMENTAL METHOD

2.1 SUBJECTS

After obtaining informed consent, eleven right-handed healthy subjects participated in the study (9 male, 2 female, mean (SD) age: 22.0 (0.77) years, range: 21-23 years). Blood hemoglobin concentration in the brain was measured used NIRStation (Shimadzu, Japan), a near-infrared imaging device. The measurement sites of NIRS are shown in Figure 1. In this report we focus on the frontal lobe, which has been identified by previous fMRI and PET studies as being involved in higher brain functions such as cognition, judgment, attention concentration and attention allocation. We analyzed channel 1 (right frontal cortex) and channel 19 (left frontal cortex) that are involved in cognition or judgment, and channel 25 (dorsolateral prefrontal cortex, near Brodmann area 46) that is involved in attention and control.

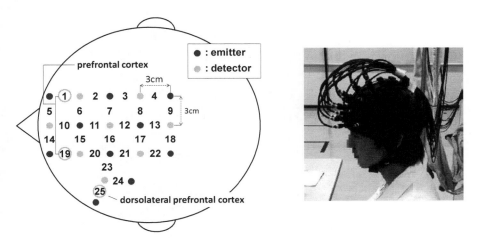

Figure 1. Schematic indicating probe mounting position (left) and a subject wearing the holder (right). Probe spacing is 3 cm. We analyzed channels 1 and 19 (prefrontal cortex) and channel 25 (dorsolateral prefrontal cortex), circled in yellow on the schematic.

2.2 TASKS

We designed two computer mousing tasks: a click task and a drag task (Figure 2). In the click task (Figure 2; left panels), the participant is required to click on the target. After each click, the position of the target changes at random. The participant is required to click on the relocated target, and the process repeats. In the drag task (Figure 2; right panels), the participant is required to click on a black square that on the left side of the screen and drag it onto a target black square on the right side of the screen. After each drag is completed, the position of the target changes at random. In both these tasks, performance is scored according to the number of successfully completed repetitions in 60 s.

We modified the difficulty of the task by changing the size of the square target. Therefore, we could decrease the usability of task by reducing the target size, or increase the usability of task by increasing the target size. In this experiment, four levels of usability were provided, with target squares of 4 pixels, 8 pixels, 16 pixels and 32 pixels. These sizes were chosen to give clear differences in the size of the target.

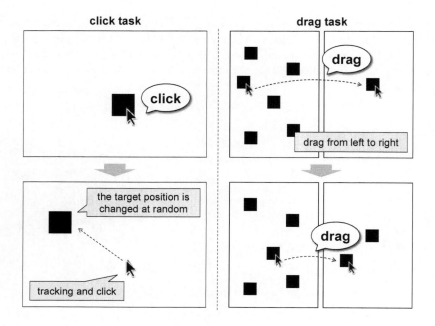

Figure 2. A schematic of the click task (left) and the drag task (right). In the click task, the target position changes random with clicking on the target. The participants were required to click on the relocated target, and this process continued for 60 s. In the drag task, participants were required to drag to the black square from the left side of the screen on to target that appeared at a random position on the right side of the screen.

2.3 EXPERIMENTAL PROTOCOL

The experiments were presented in a blocked design. The experimental protocol is shown in Figure 3. One block of trials consisted of 30 s rest (Pre), 60 s task (Task) and 30 s rest (Post); repeated five times. Before and after this 10 min block, participants sat quietly for 60 s with eyes closed and were instructed not to think about anything. Each subject performed eight blocks of trials: four blocks with the click task (4-pixel, 8-pixel, 16-pixel, and 32-pixel targets) and four blocks with the drag task (4-pixel, 8-pixel, 16-pixel, and 32-pixel targets). The two tasks were performed on separate days to prevent fatigue. The order of presentation of the two tasks was counter balanced. After completion of each 60 s task, participants completed a Japanese version NASA-TLX, a subjective evaluation of mental workload. Response time to the subjective evaluation (NASA-TLX) spend about 3 minutes.

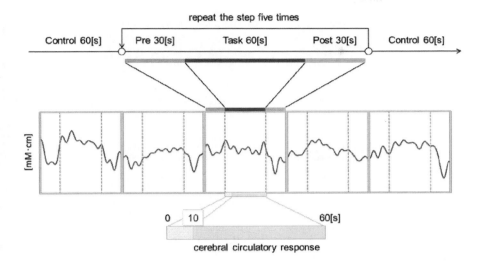

Figure 3 Schematic of the experimental design. For baseline correction, participants performed a control condition (quiet rest, eyes closed) for 60 s before and after the tasks. After the initial control condition, the following process was repeated five times: 30 s pre-task rest; 60 s task; 30 s post task rest). We used data from 10-50 s after task onset to evaluate the cerebral circulatory response.

3   ANALYSIS

With NIRS it is possible to measure oxygenated hemoglobin concentration (oxyHb), deoxygenated hemoglobin concentration and total hemoglobin concentration. In this study, we focus on analyzing oxyHb, as this is most strongly correlated with changes in regional cerebral blood flow. The frontal lobe is the most significant development site in the human brain, and is responsible for complex information processing. An increase in oxyHb indicates activation of neurons. Thus, increased oxyHb during the tasks indicates that the brain region under study had been active during the task. We interpret this as an indication that the brain region was supporting higher-order information processing.

The NIRS signal includes information that is contained in the blood, scalp and skull plates. To eliminate the effect of light absorption in other parts of the brain the NIRS signal during the control period is typically subtracted from the NIRS signal during the task period. However, this traditional evaluation method was problematic because some subjects that did not have stable oxyHb during the control period. For this reason we used a new evaluation index that was focused on changes in oxyHb over time. An example of the evaluation index is shown in Figure 4. The left panel shows the oxyHb level from a representative subject. In this example, oxyHb levels increased during the 60 s task period. In order to quantitatively assess this trend, we applied a differential filter to derive the change in oxyHb. The right panel shows the filtered data. These data show the slope of the oxyHb levels in the left panel. To evaluate the cerebral circulatory response, we summed the slope from 10 s after task onset to 10 s before task completion (shaded areas, Figure 4). Using this value, termed δ_{oxyHb}, we can assess changes in oxyHb during the task. A positive value of δ_{oxyHb} ($\delta_{oxyHb}>0$) indicates an overall increase in oxyHb level, and a negative value of δ_{oxyHb} ($\delta_{oxyHb}<0$) indicates an overall increase in oxyHb level.

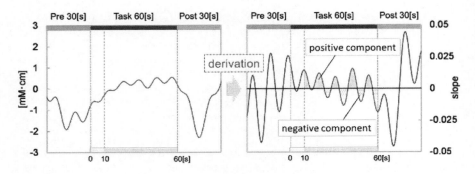

Figure 4. Representative data from one subject to illustrate the analysis procedure. The left panel shows absolute oxyHb levels throughout the three stages of the block (pre task rest, task, post task rest). The right panel shows these data after they have been passed through a differential filter, and therefore shows the change in oxyHb level. For our analysis we focus on the differentiated data. To evaluate the cerebral circulatory response induced by the task we calculated the sum of the data from 10-50 s after task onset (i.e. total positive component – total negative component). We termed this value δ_{oxyHb}.

4 RESULTS

4.1 NASA-TLX

Figure 5 shows the self-reported physical demand and frustration levels of participants, evaluated using NASA-TLX. The reduction in target size increased frustration levels in both tasks. In addition, frustration levels were higher during the drag task than during the click task. The level of perceived physical demand was higher during the drag task than during the click task, indicating that the drag task made a stronger physical demand on the participant than the click task did. In addition, levels of perceived physical demand increased when target size decreased in the click task, but not in the drag task.

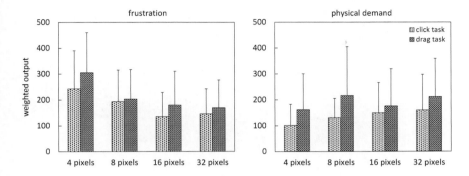

Figure 5. The left panel shows the frustration levels evaluated using NASA-TLX. The right panel shows the perceived physical demand levels evaluated using NASA-TLX. The reduction of target size increased frustration levels for both the click task and the drag task, but increased perceived physical demand only for the click task. The perceived physical demand was higher for the drag tasks than for the click tasks.

4.2 NIRS

Figure 6 shows the δ_{oxyHb} values. Significant differences were not obtained due to large inter-indicial variability. However, we obtained a similar pattern of results in both tasks. In the click task there is a negative δ_{oxyHb} value in all channels at the maximum target size of 32 pixels. In addition, at channels 1 and 19 the δ_{oxyHb} values increase as target size decreases. In channel 25 (left dorsolateral prefrontal cortex) there is a different trend; δ_{oxyHb} values decrease with as target size decreases from 16 to 4 pixels.

Similar patterns were observed in the drag task. A negative δ_{oxyHb} value at the maximum target size of 23 pixels was only observed in channel 1 (right prefrontal cortex), however, the positive δ_{oxyHb} values for channels 19 and 25 were small in magnitude. There was a tendency δ_{oxyHb} values to increase as target size decreased in channel 19 (left prefrontal cortex), however, no trend could be found for channels 1 and 25 (right prefrontal cortex and left dorsolateral prefrontal cortex).

The main finding from the NIRS data is that, in general, δ_{oxyHb} values were smallest with the maximum target size 32 pixels and largest with the minimum target size of 4 pixels.

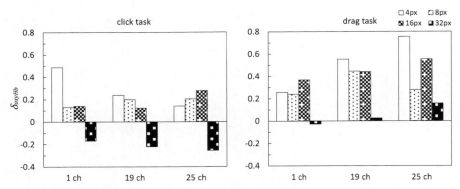

Figure 5 The left δ$_{oxyHb}$ values for the click task (left panel) and the drag task (right panel). Channel 1 represents the right prefrontal cortex, channel 19 represents the left prefrontal cortex, and channel 25 represents the left dorsolateral prefrontal cortex. In both tasks there are positive δ$_{oxyHb}$ values for all channels for the smallest target size (4 pixels). In both tasks, δ$_{oxyHb}$ values are smallest for the maximum target size (32 pixels).

5 DISCUSSION

The NASA-TLX, revealed that a reduction in target size increased frustration levels. We consider this an indication that the reduction in target size decreased usability. The reduction in target size also increased δ_{oxyHb} values, indicating an overall increase in oxyHb level. We predict that further decreases in usability would lead to further increases in oxyHb.

Negative δ_{oxyHb} values, indicative of an overall decrease in oxyHb level, were observed with a target size of 32 pixels for all channels in the click task but only for the right prefrontal cortex in the drag task. In previous studies, a decrease in oxyHb has been reported with subject fatigue and sleepiness. If the experimental tasks are simple and monotonous, as in this study, they may cause drowsiness and fatigue and therefore may be expected to reduce oxyHb. The click task is a more simple motor control task as compared with the drag task, and could therefore be expected to result in lower levels of concentration and motivation. A motivation sufficiently low to induce drowsiness may have caused the observed decrease in oxyHb levels during the click task with large target size. In the drag task it is possible to maintain a high motivation level as more complex motor control is required, and this may explain the lack of negative δ_{oxyHb} values.

The results of the subjective evaluation (NASA-TLX) indicate that the load on the motor control was higher in the drag task, and that this task was perceived to require more motion control. This is supported by the positive δ_{oxyHb} values observed during the drag task at left dorsolateral prefrontal cortex (channel 25). This area of the brain is known to be involved in attention and control, and thus an overall increase in oxyHb level can be used to support the conclusion that the drag task required high levels of motion control.

In order to eliminate the effect of the optical absorption of blood contained in the scalp and skull plates on the NIRS signal, it is common to compare the data obtained during the task to data obtained during a control period consisting of quite rest with eyes closed. However, with this analysis we found that oxyHb levels were not stable during the task, and were not consistent across subjects. One possible reason for the high inter-subject variability in hemoglobin levels during the task is high inter-subject variability during the rest period. A high level of brain activity during the intended rest of the control period would lead to apparent decreases in hemoglobin levels during the task. Additionally, in this study we used a monotonous task that could be accomplished without high levels of conscious focus. Due to the simplicity of the task, the absolute amount of increase or decrease in hemoglobin concentration is small. For these reasons, in this study we focused on changes of oxyHb during the task. We used δ_{oxyHb} as a measure of the overall change in oxyHb levels across the task. This variable has not previously been reported. The traditional analysis looking at the difference between control (resting) and task oxyHb levels is limited by the accuracy of the control measure, and the achievement of a true resting state is difficult,. Furthermore, if testing across multiple days, it is difficult to control the condition of the subject across testing sessions. Arousal state and fatigue of the participants may vary from hour to hour, and may not necessarily correspond across subjects, or within subject across testing session. We consider that the δ_{oxyHb} measure accounts for the fatigue and wakefulness levels of the participant during a particular testing session. Thus, it is possible to compensate for the shortcomings of traditional evaluation methods and it is not necessary to achieve the unification of the subject's state in the tasks and control periods. However, δ_{oxyHb} cannot be interpreted as the absolute change of hemoglobin concentration, and therefore may not be a valid measure of brain activity. Thus, to verify the usefulness of this index, while maintaining the reliability of assessment validity, there is a need to clarify the meaning of the index with conventional analysis methods.

6 CONCLUSION

We suggest the δ_{oxyHb} measure has potential when fNIRS is used to evaluate usability. However, because of high levels of inter-individual variability in our data further research is required. The validity of the δ_{oxyHb} also requires further investigation.

ACKNOWLEDGMENT

This study was supported by a research grant (Grant-in-Aid for Scientific Research (C), No. 22500415) from an independent administrative institution, the Japan Society for the Promotion of Science.

REFERENCES

Bandettini P, Wong EC, Hinks RS, et al. 1992, Time course EPI of human brain function during task activation, *Magnetic Resonance Med* 25: 390-397.

Cabeza R, Nyberg L. 2000, Imaging cognition II: An empirical review of 275 PET and fMRI studies, *J Cogn Neurosci* 12 (1): 1-47.

Haga S, Mizukami N. 1996, Japanese version of NASA Task Load Index: Sensitivity of its workload score to difficulty of three different laboratory tasks, *The Japanese journal of ergonomics* 32 (2): 71-79.

Hoshi Y, Kobayashi N, Tamura M. 2001, Interpretation of near-infrared spectroscopy signals: A study with a newly developed perfused rat brain model, *J Appl Physiol* 90: 1657-1662.

Kazunari Morimoto. 2008, Introduction of Usability Testing of Mobile Devices, *Journal of the Japan Society for Precision Engineering* 74 (2): 125-127.

Ofen N, Kao YC, Sokol-Hessner P, et al. 2007, Development of the declarative memory system in the human brain, *Nat Neurosci*, 10: 1198-1205.

R Cabeza, L Nyberg. 2000, Imaging cognition II: An empirical review of 275 PET and fMRI studies, *Journal of Cognitive Neuroscience* 12 (1): 1-47.

Sambataro F, Murty VP, Callicott JH, et al. 2008, Age-related alterations in default mode network: Impact on working memory performance, *Neurobiol Aging, in press*.

Sowell ER, Thompson PM, et al. 2004, Longitudinal mapping of cortical thickness and brain growth in normal children, *J Neurosci* 24 (38): 8223-8231.

Suda M, Fukuda M, Sato T, et al. 2009, Subjective feeling of psychological fatigue is related to decreased reactivity in ventrolateral prefrontal cortex, *Brain Research* 1252 152-160.

Suda M, Sato T, Kameyama M, et al. 2008, Decreased cortical reactivity underlies subjective daytime light sleepiness in healthy subjects: A multichannel near-infrared spectroscopy study, *Neuroscience Research* 60: 319-326.

Taoka T, Iwasaki S, Uchida H, et al. 1998, Age correlation of the time lag in signal change on EPI-fMRI, *J Computer Assisted Tomography* 22: 514-517.

Villringer A, Planck J, Hock C, et al. 1993, Near infrared spectroscopy (NIRS): A new tool to study hemodynamic changes during activation of brain function in human adults, *Neuroscience Letters* 154: 101-104.

CHAPTER 28

Team Neurodynamics: A Novel Platform for Quantifying Performance of Teams

Giby Raphael[1], Ron Stevens[2], Trysha Galloway[2], Chris Berka[1], Veasna Tan[1]

[1]Advanced Brain Monitoring, Inc.
Carlsbad, CA, USA
graphael@b-alert.com
[2]UCLA IMMEX Project
Culver City, CA, USA

ABSTRACT

This paper describes an easily deployable networked platform for collecting Electroencephalography (EEG) and Electrocardiography (ECG) from the members of a team and for analyzing and reporting the team's cognitive state in real-time. Organizations are increasingly dependent on co-located, virtual and distributed teams. Current measures of team performance depend on qualitative reporting and expert observations making it difficult to objectively equate performances across teams, training sites, or over time. Our hypothesis is that the team dynamics can be unobtrusively measured using EEG to objectively quantify the team's cognitive state. In developing this neurodynamic framework and hardware platform, we combined real-time EEG derived measures of engagement and workload from individuals into unitary metrics of team dynamics. The resulting metrics can be collected and analyzed in real-world situations, are sensitive to long and short-term task changes, and can be analyzed and reported in near real-time. The system has been implemented with tasks of varying difficulty including simple neurocognitive tasks, problem-solving decision making tasks, and complex submarine piloting and navigation tasks.

Keywords: EEG, neurophysiology, neuro synchronies, team dynamics

1 INTRODUCTION

The efficiency and performance of teams are vitally important to the success of organizations and work groups. Team oriented tasks, such as those in industry or the military rely on coherent and well-coordinated interactions as well as active participation between each member of the team. Current measures and methods for accessing team performance often rely on subjective reporting and qualitative observations. Moreover, they rely on externalized events such as the composition of the team and the characteristics of their work. The study of teamwork based on more internalized measures such as team cognition involves analyzing interdependent acts of the individual activity directed towards achieving a collective goal. As members of a team perform their duties each will exhibit varying degrees of cognitive components such as attention, workload, engagement, etc. and the levels of these components at any one time will depend (at least) on 1) what that person was doing at a particular time, 2) the progress the team has made toward the task goal, and 3) the composition and experience of the team.

Electroencephalography (EEG) has traditionally been viewed as a tool for studying individual cognition in the milliseconds to seconds range. Recently we have begun applying these technologies to the study of the cognitive organization of teams and have developed a neurophysiologic framework for studying complex teamwork called Team Neurodynamics (NDs) with an analytic protocol for measuring Neurophysiologic Synchronies (NS). NS are low level data streams defined by the second-by-second quantitative co-expression of the same neurophysiologic / cognitive measure by different team members. Analysis of NS expressions provides real-time insights into team cognition.

NS based teaming metrics have been studied across multiple teams and tasks including simple neurocognitive tasks, educational problem solving tasks, and more complex simulated Submarine Piloting and Navigation (SPAN) tasks involving Junior Officers in the Submarine Officer Advanced Course (SOAC) at the Submarine Learning Center (SLC). Consistent patterns were found across the tasks that distinguished team performance. We believe that the measures can be applied to objectively assess team interactions for within and across team comparisons and to adapt and optimize team training. Further development of these measures can be used to train models that can predict team performance in real-time.

An easily deployable networked platform has been developed in order to collect EEG and ECG from individual members of a team and has been used to analyze and report performance in real-time. Advanced Brain Monitoring's (ABM) highly portable and unobtrusive wireless EEG device enables collection of data without obstructing the teaming tasks. The data from the individuals are transmitted through the network and collected at a central server for processing and display. The platform has a layered architecture such that it can be deployed across various teaming environments with minimal overhead.

2 METHODS

2.1 Analysis of EEG

ABM's B-Alert® system contains an easily-applied wireless EEG headset that includes intelligent software designed to identify and eliminate multiple sources of biological and environmental contamination and allows for real-time classification of cognitive state changes even in challenging environments. The 10-channel wireless headset include sensor site locations according to the International 10-20 system: F3, F4, C3, C4, P3, P4, Fz, Cz, POz; monopolar referenced to linked mastoids and ECG. The ABM B-Alert® software acquires the data and quantifies alertness, engagement and mental workload in real-time (Berka, Levendowski et al. 2004; Poythress, Russell et al. 2006; Berka, Levendowski et al. 2007).

The data processing begins with the detection and decontamination of artifacts such as eye-blinks, EMG, etc. from raw EEG and subsequent computation of power spectral densities (PSD) of the filtered data. The PSDs from specific EEG channels are then used by a generative classifier to output second-by-second probabilities of cognitive states such as: Distraction, Low-Engagement, High-Engagement (EEG-E), High-EEG-Workload (EEG-WL) (Levendowski, Berka et al. 2001; Berka, Levendowski et al. 2004; Berka, Levendowski et al. 2007).

EEG-WL metric reflects working memory load, problem solving, integration of information, and analytical reasoning. The EEG-E metric reflects information-gathering, visual processing, allocation of attention. It shares similarities with alertness or attention and can be driven by visual, tactile, and / or auditory processing (Kahol, French et al. 2006; Stevens, Galloway et al. 2007; Stevens, Galloway et al. 2007; Stevens, Galloway et al. 2007; Berka 2008). It is analogous to the EEG-rhythm-based attention measures that are often associated with alpha power dynamics (Jung and et al. 1997; Kelly, Docktree et al. 2003; Huang, Jung et al. 2007). Operationally, precise cognitive terms will be difficult to associate with EEG-derived measures of cognition in the context of teamwork, and functional associations will need to be derived empirically.

2.2 Computation of Team Neurodynamics

A modeling approach was developed using the NS normalized second-by-second values of EEG-E concatenated into a vector representing the levels being expressed by each team member (Figure 1A) (Stevens, Galloway et al. 2011). Using artificial neural network (ANN) technologies, these vectors were modeled onto collective team variables that are termed neurophysiologic synchronies of engagement (NS_E) and workload (NS_WL). NS_E and NS_WL capture the engagement or workload of each of 6 team members as well as of the team as a whole (Stevens, Galloway et al. 2012). ANN classification of these second-by-second vectors create a symbolic state space that shows the possible combinations of Engagement or Workload across members of the team (Figure 1B).

The frequency counts for each node (NS) represent the Team Neurodynamics (NDs) across a session. The NDs for a team are identified and then analyzed within and across a session: measuring the change, frequency of occurrence, and variability in each ND as shown in Figure 1C. The NDs can be presented in real-time as the second-by-second quantitative co-expressions of the same EEG state by different members of the team.

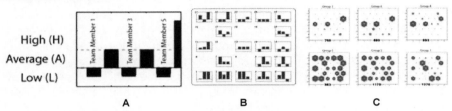

Figure 1 (*A*) Symbolic representation of EEG-E of 3-person team. (*B*) Neural Network pattern. (*C*) ND-frequency maps. Frequency is proportional to the filling in the hexagons.

2.3 Architecture of the Team Neurodynamics Platform

The NDs platform provides a streamlined solution to collect and analyze data concurrently from multiple teams and individuals. A maximum of 3 teams are supported simultaneously with each team limited to a maximum of six individuals. The platform was designed to be highly modular such that it can be configured and applied to extract teaming metrics in generic contexts. The platform is based on a client-server-architecture and utilizes standard networks for data distribution. The computational load is distributed across the network; individual metrics are computed locally while the teaming metrics are computed in the central server.

Each player-module runs on individual computers and the data from ABM's wireless EEG devices is collected and processed locally. A central administrator-module was developed for data aggregation and for running algorithms related to the teaming metrics. All modules were developed as standalone applications and host an internal server as well as client sub-modules in order to communicate with other modules in the network as shown in Figure 2.

The player-module acquires data from the device, filters the data using proprietary artifact-decontamination algorithms, computes power spectral densities, and classifies the data using ABM's proprietary machine-learning algorithms. The raw and processed data are then stored locally and also made available to external clients via its internal TCP server. The Data-Aggregation client in the administrator-module collects data from the TCP servers of all active player-modules, for collective processing and display. In order to support dynamic-joining of each player to the network, a UDP server with a universal ID-port was implemented in the administrator-module. Each player-module was required to introduce itself via its ID-Client and send all its connection details to the administrator-module prior to joining the network.

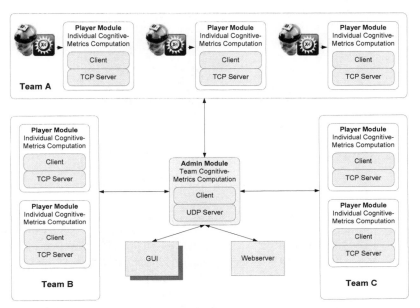

Figure 2 Block diagram of Team Neurodynamics Platform. The modules communicate via dedicated clients and servers. The player modules compute individual cognitive metrics and the Admin-module compute teaming metrics.

In order to allow access to the modules programmatically (for coupling third-party applications), they were developed as a Software Development Kit (SDK), with well-defined Application Programming Interfaces (APIs). The administrator-module was coupled with a dedicated web-server for optional remote monitoring and control of the teaming tasks, via a web-browser. The NDs metrics are computed in the administer-module in real-time from the aggregated data. An elaborate Graphical User Interface (GUI) is interfaced to the administrator-module in order to visualize both within and across team parameters. Network Time Protocol (NTP) is used to maintain timing synchronization between the modules.

3   RESULTS

3.1   Simple Neurocognitive Tasks

Individual EEG-Workload levels were measured during: 1) Baseline neurocognitive assessment task (3 Choice- Vigilance Test), 2) Individual problem solving exercises, and 3) Team problem solving exercises. It was found that EEG-Workload measures increased during team tasks, but time to solve the problem was greatly reduced (Stevens, Galloway et al. 2009). Thus real-time monitoring of EEG-WL is useful for optimizing team performance. The results supplements research efforts using non-physiological workload metrics to access team cognition (Urban, Bowers et al. 1995; Moye and Langfred 2005).

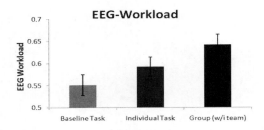

Figure 3 EEG-Workload measures during 3 tasks. Workload increased during team tasks, but time to solve problem was greatly reduced.

3.2 Educational Tasks

In this task, teams of three high school students explored an online IMMEX™ problem space where the goal was to make a decision of whether the simulated person should seek help for substance abuse. One member of the team accessed physiologic and neurophysiologic data, one member examined social issues such as school / job performance, difficulties with the law, interactions with peers, etc., and the third person led the group interactions and guided the decision. Comparison of three slowest teams with three fastest teams showed that the variability of Team ND patterns were highly related to the time to solve ($R^2 = 0.5713$) (Stevens, Galloway et al. 2009). This indicates that efficient teams show more focused (less distributed) ND patterns.

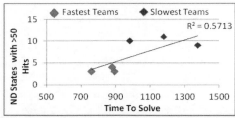

Figure 4 Variability of ND states vs. Performance. It was seen that fastest teams had more focused (less distributed) ND patterns.

3.3 Military Training Tasks

The application of the approach was tested on a very complex training task which involved the safe piloting of a submarine. Junior officers in the SOAC at the SLC performed the tasks in the SPAN, a high fidelity training simulator. The dynamically programmed events include encounters with approaching ship traffic, the need to avoid nearby shoals, changing weather conditions, and instrument failure. There are also task-oriented cues to provide information to guide the mission, and team-member cues that provide information on how other members of the team are performing / communicating. Finally, there are adaptive behaviors that help the team adjust in cases where one or more members are under stress or are not familiar with aspects of the unfolding situation.

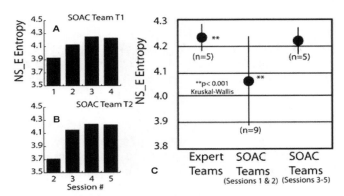

Figure 5 Changes in the entropy of Engagement NS patterns as Junior Officers in the SOAC gain navigation experience. (A,B) The overall entropy of NS_E data streams during the Scenario portions of SPAN (C) subsequent performances by SOAC teams and comparison with 5 experienced submarine navigation teams (SUB).

For this particular task we looked at a measure of entropy (Shannon 1951) to describe the level of uncertainty in NS data streams. Entropy is a measure derived from the string of NS nodes chronologically expressed throughout a teaming session. Entropy represents the "spread" or "amount of mix" of NS nodes over a specified time epoch. High entropy indicates more distributed NS patterns, and low entropy indicates more focused (less distributed) NS patterns over a given epoch.

Figure 5 shows entropy values of several SPAN sessions. As the SOAC teams (Team 1 & Team 2) performed multiple SPAN sessions, the level of entropy approaches that of an expert team (SUB). This indicates that higher entropy (i.e., exhibiting a more distributed repertoire of NS patterns, rather than locking into fewer NS patterns) is a characteristic of expertise in the SPAN environment. Furthermore, team NS patterns can track learning trajectories as novices acquire expertise.

4 DISCUSSION

The formation and functioning of highly flexible and adaptive team collaborations is essential across industry, sports, academia and in the military. One broad challenge facing all forms of teamwork is the measurement of team interactions to assess the quality of performance and decision-making early enough to intervene in order to optimize the team (Cohen 1997; Salas, Cooke et al. 2008). This applies to co-located teams, virtual teams and networked, hierarchically structured teams. Meeting this challenge requires the development of objective, unobtrusive and real-time measures of team performance that can be practically implemented in ecologically valid training simulations and real world environments.

Measurement of team performance is generally conducted by "expert observers" and/or subjective reports from individual team members usually after the teaming exercise is complete. The expert assessment approach relies on the external observables (team performance, communication and errors) with no ability to discern the underlying motivations or cognitive processes of teams. In addition, the expert observer may introduce biases based on prior experience. Subjective reports may reveal additional information on the internal states of the team members but are limited by the ability and willingness of the individuals to articulate their experience and by the delayed timing of the debrief from the action of the teaming scenario. Self-reports are also well documented to be limited in variability, and often fail to accurately reflect the metrics of interest, due to reporting bias (Pryor, Gibbons et al. 1977). Improved methods and objective metrics for understanding the factors that lead to the formation of productive and cohesive teams will enhance team training.

The Team Neurodynamics platform is being designed to assess multi-echelon team interactions and to provide a window on the formation of "team cognition". One objective is to expand the concept of individual cognitive readiness to team cognitive readiness by developing an approach to mapping a Team Mental Model (TMM), defined as team members' shared, organized understanding and mental representation of knowledge about key elements of team performance (Cooke, Stout et al. 1997; Mathieu, Heffner et al. 2000). Shared TMMs reportedly enhance team performance by allowing team members to anticipate the needs of the team and distribute task loads accordingly (Stout, Cannon-Bowers et al. 1996).

The construct of metacognition or "team cognition" may provide a framework for the development of a model for assessment of teams. This conceptual understanding of team interactions suggests a model wherein team members pool knowledge and create shared mental models to achieve successful team performance. For instance, teams that are in high concordance should require lower metacognitive reflection and monitoring whereas teams that are not collaborating or failing to properly distribute effort across team members would require extensive metacognitive monitoring of the environment and team actions. These associations arise from the need to recursively monitor and model the behavior of other agents involved in the team situation. Mapping the EEG-based metrics of teams offer the potential of monitoring the metacognitive state in real-time. Interventions can then be made early in the team interaction before deleterious patterns such as groupthink or failed communications sabotage the team performance.

5 CONCLUSIONS

The NDs platform presented in this paper can be applied to multiple contexts to unobtrusively measure team performance in real-time. Three applications of varying difficulties were presented, one from a research perspective, one from an educational perspective, and one from a complex training perspective. The results indicate that the analysis based on Team NS can provide objective real-time insights into team cognition in order to 1) distinguish between expert and novice teams, 2) track learning trajectories as novices gain expertise, and 3) show state changes associated with key task events within a training session. The modular and easy to use platform can be quickly deployed in order to determine the functional status of a team, assess the quality of the teams' performance and decisions, and to adaptively rearrange the team or task components to better optimize the team. The platform can be easily interfaced with other third party applications via well defined programmable interfaces. It is currently used in other applications such as studying team interaction in goal oriented military training scenarios, decision making in business contexts, and developing novel approaches to improve team interactions.

ACKNOWLEDGMENTS

This work was supported in part by The Defense Advanced Research Projects Agency (DARPA) under contract number(s) NBCHC070101, NBCHC090054, and NSF SBIR grant IIP 0822020. The views, opinions, and/or findings contained are those of the authors and should not be interpreted as representing the official views or policies, either expressed or implied, of the Defense Advanced Research Projects Agency or the Department of Defense.

REFERENCES

Berka, C. (2008). Assessing Students Mental Representatives of Complex Problem Spaces with EEG Technologies. Human Factors and Ergonomics Society 52nd Annual Meeting. New York, NY.

Berka, C., D. Levendowski, et al. (2004). "Real-time Analysis of EEG Indices of Alertness, Cognition, and Memory Acquired with a Wireless EEG Headset." Int J Hum Comput Interact **17**(2): 151-170.

Berka, C., D. Levendowski, et al. (2007). "EEG Correlates of Task Engagement and Mental Workload in Vigilance, Learning and Memory Tasks." Aviation Space and Environmental Medicine **78**(5): B231-B244.

Cohen, S. B., DE (1997). "What makes teams work: group effectivenss research from the shop floor to the executive suite." Journal of Management **23**(3): 239-290.

Cooke, N., R. Stout, et al. (1997). "Broadening the Measurement of Situation Awareness through Cognitive Engineering Methods." Proceedings of the Human Factors and Ergonomics Society Annual Meeting **1**: 215-219.

Huang, R. S., T. P. Jung, et al. (2007). Multi-scale EEG brain dynamics during sustained attention tasks. IEEE International Conference on Acoustics, Speech, and Signal Processing. Honolulu, Hawaii. **4**: 1173-1176.

Jung, T. P. and et al. (1997). "Estimating alertness from the EEG power spectrum." IEEE Trans Biomed Eng. **44**(1): 60-69.

Kahol, K., J. French, et al. (2006). Evaluating the Role of Visio-Haptic Feedback in Multimodal Interfaces through EEG Analysis. Augmented Cognition: Past, Present and Future. D. Schmorrow, K. Stanney and L. Reeves. Arlington, VA, Strategic Analysis, Inc.: 289-296.

Kelly, S. P., P. Docktree, et al. (2003). EEG Alpha power and coherence time courses in a sustained attention task. International Conference on Neural Engineering: 83-86.

Levendowski, D., C. Berka, et al. (2001). "Electroencephalographic indices predict future vulnerability to fatigue induced by sleep deprivation." Sleep **24**(Abstract Supplement): A243-A244.

Mathieu, J., T. Heffner, et al. (2000). "The influence of shared mental models on team process and performance." Journal of Applied Psychology **85**(2): 273-283.

Moye, N. and C. Langfred (2005). "Information sharing and group conflict: going beyond decision making to undestand the effects of information sharing on group performance." International Journal of Conflict Management **15**(4): 381-410.

Poythress, M., C. Russell, et al. (2006). Correlation between Expected Workload and EEG Indices of Cognitive Workload and Task Engagement. Augmented Cognition: Past, Present and Future. D. Schmorrow, K. Stanney and L. Reeves. Arlington, VA, Strategic Analysis, Inc.**:** 32-44.

Pryor, J. B., F. X. Gibbons, et al. (1977). "Self-focused attention and self-report validity." Journal of Personality **45**(4): 513-527.

Salas, E., N. Cooke, et al. (2008). "On Teams, Teamwork, and Team Performance: Discoveries and Developments." Human Factors: the journal of the HUman Factors and Ergonomics Society **50**(3): 540-547.

Shannon, C. E. (1951). "Prediction and entropy of printed English." The Bell System Technical Journal **30**: 50-64.

Stevens, R., T. Galloway, et al. (2007). Allocation of Time, EEG-Engagement and EEG-Workload Resources as Scientific Problem Solving Skills are Acquired in the Classroom. Augmented Cognition: Past, Present, & Future. D. Schmorrow, K. Stanney and L. Reeves. Arlington, VA, Strategic Analysis, Inc.

Stevens, R., T. Galloway, et al. (2007). EEG related Changes in Cognitive Workload, Engagement, and S. User Modeling Conference. Athens, Greece.

Stevens, R., T. Galloway, et al., Eds. (2007). Exploring Neural Trajectories of Scientific Problem Solving Skill Acquisition. Augmented Cognition: Past, Present, & Future. Arlington, VA, Springer-Verlag Berlin Heidelberg.

Stevens, R., T. Galloway, et al. (2011). Linking models of team neurophysiologic synchronies for engagement and workload with measures of team communication. 20th Conference on Behavior Representation in Modeling and Simulation (BRIMS), Sundance, UT.

Stevens, R., T. Galloway, et al. (2009). Neurophysiologic Collaboration Patterns During Team Problem Solving. Human Factors and Ergonomics Society (HFES), San Antonio, TX.

Stevens, R., T. Galloway, et al. (2012). "Cognitive neurophysiologic synchronies: What can they contribute to the study of teamwork?" Human Factors: in Press.

Stout, R., J. Cannon-Bowers, et al. (1996). "The role of shared mental models in develping team situational awareness: implications for training. ." Training Research Journ **2**: 86-116.

Urban, J., C. Bowers, et al. (1995). "Workload, team structure, and communication in team performance." Military Psychology **7**(2): 123-139.

Section VII

Neuroergonomics: Applications

CHAPTER 29

Challenges and Opportunities for Inserting Neuroscience Technologies into Training Systems

Roy Stripling

Center for Research on Evaluation, Standards, and Student Testing (CRESST)
University of California, Los Angeles (UCLA)
Los Angeles, CA
roy.stripling@cse.ucla.edu

ABSTRACT

Legitimate dry-contract electrode Electroencephalogram (EEG) systems finally seem to be within reach. This technical advance opens possibilities for innovation in student/trainee assessment by reducing one of the long-standing barriers to EEG use in the classroom/simulation lab environments: user acceptance. However, three major barriers remain. First, motion, muscle, and ambient electrical artifacts are still a problem. Supervised post-processing by experienced EEG technicians is still common practice. Classrooms and simulation labs do not and will not have the benefit of skilled technicians. Automated methods for reliably cleaning the data, without loss of information, are necessary. Second, the pedagogical value of EEG-based assessment must be convincingly demonstrated. Most efforts to enhance training or education through application of EEG-based assessment have been solutions in search of problems. Instead of asking what EEG can tell us and how we can use that knowledge to enhance training and education, we need to ask what the high priority training and education gaps are and whether EEG-based assessments can help fill them. Finally, signal variation, both between individuals and within individuals from day-to-day, are frequent problems in EEG signal detection. In the research environment, some of this variation is minimized through practices that will not be acceptable in the mainstream environment (e.g., limiting head movement and screening out left-handers). This variation in signal source detection needs flexible and robust solutions so that all students/trainees can receive any benefits that are brought about through the introduction of EEG-based assessment practices. This paper discusses these barriers and suggests solutions that are necessary for the long-heralded brain-based education and training revolution to take hold.

Keywords: Electroencephalogram, EEG, Cognitive Neuroscience, Education, Assessment

1 INTRODUCTION

School systems perennially work under tight budgets to meet increasing standards and requirements with overloaded teachers (Mayers, 2010). This observation appears to hold true across the realm of education environments (k-12, adult education, military training). In all cases, training and education opportunities compete with other pressing needs for limited resources, and practical compromises have to be made. Educators and trainers may look to the research community for help, but real solutions are slow to transfer (Gersten & Dimino, 2001).

For years, cognitive neuroscientists have envisioned fielding practical applications of their technologies, most notably electroencephalography (EEG) in classroom or computer based learning environments (e.g., Gevins, Long, Sam-Vargas, & Smith, 1995; Smith, McEvoy, & Gevins, 1999; Berka et al., 2007; Coyne, Baldwin, Cole, Sibley, & Roberts, 2009; Craven, Tremoulet, Barton, Tourville, & Dahan-Marks, 2009). This body of work and related research reflects attempts to use EEG-based systems to assess such things as student drowsiness, engagement, cognitive workload, and student attention or distraction. Yet, no EEG or other brain activity monitoring technology has successfully transitioned to any real-world learning environment. Four likely reasons for this failure are that these technologies and applications are too fragile, too expensive, too difficult to work with, and pedagogically irrelevant or ill-conceived.

Recent technological innovations appear to be on the verge of resolving the first two of these barriers. Emotiv Systems (emotiv.com) offers a low-cost and apparently reasonably durable EEG system. Also of importance, several research groups have developed, and at least one commercial entity is selling, dry contact EEG electrode systems (e.g., Chi, Jung, & Cauwenberghs, 2010; Liao et al., 2012; Liao, Wang, Chen, Chang, & Lin, 2011; Fiedler et al., 2010; Gargiulo et al., 2010; g.tec (gtec.com)). If effective, these electrodes, which do not require skin preparation or the use of wet conductive agents, could make EEG systems much more user-palatable and friendly. However, reducing cost, increasing durability, and improving the quality of user experience through dry contact electrodes is still not sufficient to foster acceptance and adoption in educational settings. Substantial progress needs to be made in the areas of automated artifact processing, pedagogical relevance, and in the development of flexible cognitive state detection algorithms.

2 AUTOMATED ARTIFACT PROCESSING

It is no secret that brain-sensing technologies are subject to contamination from a wide range of sources (see Corby & Kopell, 1972; Matsuo, Peters, & Reilly, 1975; Delorme, Sejnowski, & Makeig, 2007). Chief among these are motion and muscle artifacts. In the lab, these artifacts are routinely dealt with by optimizing electrode impedance, instructing participants to relax and limit unnecessary movements, and designing tasks that minimize eye and head movements. These tricks of the trade will not suffice for EEG applications in more dynamic education settings. If used only with computer-based education and training methods, one could hope for a reduced set of head and eye movements relative to other classroom activities. However, efforts to further constrain head and eye movements within a computer-based educational task would limit the operational field of view to a small area of the screen, which in turn, would limit the variety, variability, and complexity of tasks that could be used. Furthermore, in educational settings, it is reasonable to assume that recommendations to minimize facial muscle activity (tension in jaw muscles, brow furrowing, etc.) will largely go unheeded.

Fortunately, researchers have sought effective and efficient automated methods for removing eye-blink, eye movement, and muscle artifacts for more than 30 years (e.g., Whitton, Lue, & Moldofsky, 1978; Woestenburg, Verbaten, & Slangen, 1983; Croft & Barry, 2000; Klados, Papadelis, Braun, & Bamidis, 2011; Priyadharsini & Rajan, 2012; Hsu, Lin, Hsu, Chen, & Chen, 2012). Reviewing and critiquing the merits of this extensive body of work is beyond the scope of this paper. However, it is worth noting that much of this work has been focused on removing artifacts observed in lab and clinical environments. Educational users (students and teachers) cannot be expected to have the knowledge and experience to recognize and correct artifact contamination problems. Any system they use will need to be capable of self-diagnosis and be able to provide step-by-step directions to teachers on how to identify and replace worn or damaged parts when software-based corrections are insufficient to clean or gracefully exclude contaminated data from ongoing assessments of students' EEG data. Achieving success in delivering EEG-based cognitive neuroscience systems to educational and training environments will therefore require developing very robust artifact correction algorithms that are embedded in intelligent, self-assessing software.

3 PEDAGOGICAL NEEDS

Assuming that signal processing software achieves the necessary level of sophistication to support EEG-based assessments and interventions in educational and training settings, what then? Does the cognitive neuroscience community have solutions that are of interest to education and training practitioners?

It has been argued that the cognitive neuroscience community, with its roots based in experimental psychology, neuropsychology, and neuroscience, is poorly oriented towards meeting the needs of educators and trainers (Davis, 2004; McNamara, 2006; Samuels, 2009). Its traditional focus on fundamental questions related to how the brain gives rise to the mind (where in the brain mental functions are carried out), as well as its reliance on relatively stringent and isolated experimental practices, render it more relevant to clinical psychiatry and neurosurgery than to education and training (Szűcs & Goswami, 2007). However, as the field of cognitive neuroscience has matured, the desire and effort to extend its fundamental findings to applied domains such as education and training has grown (McNamara, 2006; Szűcs & Goswami, 2007; Samuels, 2009). Over the past decade, both traditional neuroscience groups, such as the Society for Neuroscience (SFN), and new groups, such as the International Mind, Brain, and Education Society (IMBES), have begun to address this challenge directly. Further, many cognitive neuroscience researchers have also attempted to apply their methods, knowledge, and technologies to the task (e.g., Stripling, Becker, & Cohn, 2005; Berka et al., 2007; Luu, Tucker, & Stripling, 2007; Coyne et al., 2009; Craven et al., 2009; Carroll, Fuchs, Hale, Dargue, & Buck, 2010; Campbell, Belz, Scott, & Luu, 2011). But if these researchers, like the field of cognitive neuroscience itself, have their roots in experimental psychology, neuropsychology, and neuroscience, then how well will their hypotheses, insights, and conclusions map to the needs of the real-world educator/trainer? What pedagogical relevance will their contributions have to educators and trainers? According to Samuels (2009), McNamara (2006) "found that journal articles in cognitive science often claimed educational implications, but results of these studies were rarely considered from the perspective of educators, nor did researchers attempt to include educators in either creating research projects or conducting analyses." Samuels goes on to ask, "If cognitive psychology itself demonstrates few strong collaborations with education, how can it act as a mediator for neuroscience in education?"

So what, if anything, do educators and trainers actually want or need to learn from cognitive neuroscientists? There have been very few attempts to answer this question in the public domain. Pickering and Howard-Jones (2007) conducted a survey of 189 educators, most of whom were attending either the Education and Brain Research conference or the Learning Brain Europe conference in 2005. This study was focused more on what kinds and quality of exposure to brain research findings educators were getting rather than on what educators wanted and needed from brain researchers. However, the authors drew two general conclusions relevant to this question. They found that the educators in their study have an "enthusiasm for a role of neuroscience in education" and that "an understanding of the workings of the brain was seen as import in the design and delivery of educational programs for children." In 2009, the Society for Neuroscience (SFN) held a meeting entitled "Neuroscience Research in Education Summit" (SFN Report, 2009). However, if the published list of attendees accurately reflects participation, the entire education community was represented by just four pre-k-12 grade education administrators.

In short, the cognitive neuroscience community remains very much in the dark with regards to what teachers and trainers want and need to know. What is needed is a detailed and comprehensive survey to determine the needs, desires, and priorities of educators and trainers with regards to brain science research. This survey should be sent to educators and trainers serving all socio-economic groups and working in rural, suburban, and urban districts. In addition to answering directed questions on this topic, respondent descriptive data should be collected including, but not limited to, what subjects and grade-levels they teach, how many years of experience they have, and what kinds of exposure they have had to brain sciences (ideally identifying common misconceptions within the population), what kinds of schools they teach in (e.g., public, private, charter), etc. Follow up interviews with a smaller, but representative set of teachers and trainers should further explore teaching conditions, resource availability, and attempt to gauge temperament towards introducing different types of cognitive neuroscience-based instructional methods in the classroom.

4 FLEXIBLE SOLUTIONS FOR REAL-WORLD VARIABILITY

Even if they are equipped with robust systems that deal with artifacts and trouble-shooting and have an understanding of what the needs and capabilities of educators/trainers are, applied cognitive neuroscientists still will not be able deliver useful EEG-empowered technologies to the education and training community. The knowledge base of cognitive neuroscience is still too narrowly focused on stringent and isolated laboratory tasks and settings that do not support the immediate leap to the complex, dynamic, and social environment of educational and training settings. Most cognitive neuroscience research seeks to increase statistical power and validity by eliminating common sources of variability found in the real-world (e.g., second language speakers, left-handed people, and age ranges not found in the typical university student population). Even the best EEG-based cognitive state assessment algorithms are typically developed for application in real-world settings based on data collected from only hundreds of individuals performing a handful of tasks in controlled settings (Berka et al., 2004, 2007; Davis, Popovic, Johnson, Berka, & Mitrovic, 2009; Jung, Makeig, Stensmo, & Sejnowski, 1997; Wilson & Fisher, 1995). This type of work is an essential first step, but is not sufficient by itself. Armed with these lab-validated algorithms as initial hypotheses, a broad and intensive effort to collect and analyze EEG data from thousands, or tens of thousands of individuals "in the field" is needed. This data should be collected in real education and training settings with participants performing both standard tasks (used to derive the initial cognitive state assessment algorithms) and a much broader set of real-world education and training tasks. The participant population should be representative of the true variability of the student population with respect to age, gender, first-language, handedness, ethnic background, socio-economic status, etc.

The cost of such an undertaking is substantial and would likely need to be distributed across multiple institutions. However, the scale could also drive down equipment costs by creating a larger education and training market for rugged, robust, self-assessing EEG systems. Such distributed, large-scale research efforts are carried out in other fields of research by leveraging Internet infrastructures and crowd sourcing principles (Andersen et al., 2012; Eiben et al., 2012; Khatib et al., 2011; Savage, 2012). These approaches seek to maintain statistical power and validity by trading massive sample size for isolation of error sources. As a relatively new methodology, it may carry unknown risks to valid implementation, especially when putting the data collection tools in the hands of unsupervised participants. However, the output of such an undertaking would be a rich repository of annotated, task-referenced brain activity data collected from tens of thousands of people carrying out real-world activities. This repository itself would be a unique resource for the cognitive neuroscience community and would serve as a natural and robust fulcrum for moving basic laboratory research into applied, real-world settings and domains.

5   CONCLUSIONS

Cognitive neuroscientists interested in applying their findings and EEG-based technologies to real-world environments are inching closer to this possibility with the introduction of dry-contact EEG electrodes and low-cost, fairly rugged EEG systems. However, substantial progress must be made in three notable areas. First, automated artifact removal algorithms need to be powerful enough to operate in completely unsupervised and uncontrolled, real-world environments. Second, the cognitive neuroscience community needs an education in educator and trainer needs, priorities, and practical limitations. Third, an ambitious effort to leverage crowd sourcing approaches to real-world EEG data collection and analysis is needed. While such an endeavor would be challenging and costly, it would create a new and unique resource for cognitive neuroscientists in the form of a rich repository of annotated, real-world brain activity data that would serve as a critical enabler, moving cognitive neuroscience findings out of the lab and into the real-world.

REFERENCES

Andersen, E., O'Rourke, E., Liu, Y.-E., Snider, R., Lowdermilk, J., Truong, D., Cooper, S., et al. (2012). The Impact of Tutorials on Games of Varying Complexity. Presented at the CHI'12.

Berka, C., Levendowski, D. J., Cvetinovic, M. M., Petrovic, M. M., Davis, G., Lumicao, M. N., Zivkovic, V. T., et al. (2004). Real-Time Analysis of EEG Indexes of Alertness, Cognition, and Memory Acquired With a Wireless EEG Headset. *International Journal of Human-Computer Interaction*, *17*(2), 151–170. doi:10.1207/s15327590ijhc1702_3

Berka, C., Levendowski, D. J., Lumicao, M. N., Yau, A., Davis, G., Zivkovic, V. T., Olmstead, R. E., et al. (2007). EEG Correlates of Task Engagement and Mental Workload in Vigilance, Learning, and Memory Tasks. *Aviation, Space, and Environmental Medicine*, *78*(Supplement 1), B231–B244.

Campbell, G. E., Belz, C. L., Scott, C. P. R., & Luu, P. (2011). EEG Knows Best: Predicting Future Performance Problems for Targeted Training. In D. D. Schmorrow & C. M. Fidopiastis (Eds.), *Foundations of Augmented Cognition. Directing the Future of Adaptive Systems* (Vol. 6780, pp. 131–136). Berlin, Heidelberg: Springer Berlin Heidelberg. Retrieved from http://www.springerlink.com/content/6335375322116171/

Carroll, M., Fuchs, S., Hale, K., Dargue, B., & Buck, B. (2010). Advanced Training Evaluation System: Leveraging Neuro-Physiological Measurement to Individualize Training (Vol. 2010). Presented at the The Interservice/Industry Training, Simulation & Education Conference (I/ITSEC). Retrieved from http://ntsa.metapress.com/link.asp?id=f822052381327663

Chi, Y. M., Jung, T.P. & Cauwenberghs, G. (2010). Dry-contact and noncontact biopotential electrodes: methodological review. *IEEE Reviews in Biomedical Engineering, 3*, 106–119. doi:10.1109/RBME.2010.2084078

Corby, J. C., & Kopell, B. S. (1972). Differential Contributions of Blinks and Vertical Eye Movements as Artifacts in EEG Recording. *Psychophysiology, 9*(6), 640–644. doi:10.1111/j.1469-8986.1972.tb00776.x

Coyne, J. T., Baldwin, C., Cole, A., Sibley, C., & Roberts, D. M. (2009). Applying Real Time Physiological Measures of Cognitive Load to Improve Training. In D. D. Schmorrow, I. V. Estabrooke, & M. Grootjen (Eds.), *Foundations of Augmented Cognition. Neuroergonomics and Operational Neuroscience* (Vol. 5638, pp. 469–478). Berlin, Heidelberg: Springer Berlin Heidelberg. Retrieved from http://www.springerlink.com/content/443wgk24763r3531/

Craven, P. L., Tremoulet, P. D., Barton, J. H., Tourville, S. J., & Dahan-Marks, Y. (2009). Evaluating Training with Cognitive State Sensing Technology. In D. D. Schmorrow, I. V. Estabrooke, & M. Grootjen (Eds.), *Foundations of Augmented Cognition. Neuroergonomics and Operational Neuroscience* (Vol. 5638, pp. 585–594). Berlin, Heidelberg: Springer Berlin Heidelberg. Retrieved from http://www.springerlink.com/content/670p48hu1m5202r3/

Croft, R. J., & Barry, R. J. (2000). Removal of ocular artifact from the EEG: a review. *Neurophysiologie Clinique/Clinical Neurophysiology, 30*(1), 5–19. doi:10.1016/S0987-7053(00)00055-1

Davis, A. (2004). The Credentials of Brain-Based Learning. *Journal of Philosophy of Education, 38*(1), 21–36. doi:10.1111/j.0309-8249.2004.00361.x

Davis, G., Popovic, D., Johnson, R. R., Berka, C., & Mitrovic, M. (2009). Building Dependable EEG Classifiers for the Real World – It's Not Just about the Hardware. In D. D. Schmorrow, I. V. Estabrooke, & M. Grootjen (Eds.), *Foundations of Augmented Cognition. Neuroergonomics and Operational Neuroscience* (Vol. 5638, pp. 355–364). Berlin, Heidelberg: Springer Berlin Heidelberg. Retrieved from http://www.springerlink.com/content/8p57134n53q3r303/

Delorme, A., Sejnowski, T., & Makeig, S. (2007). Enhanced detection of artifacts in EEG data using higher-order statistics and independent component analysis. *NeuroImage, 34*(4), 1443–1449. doi:10.1016/j.neuroimage.2006.11.004

Eiben, C. B., Siegel, J. B., Bale, J. B., Cooper, S., Khatib, F., Shen, B. W., Players, F., et al. (2012). Increased Diels-Alderase activity through backbone remodeling guided by Foldit players. *Nature Biotechnology, 30*(2), 190–192. doi:10.1038/nbt.2109

Fiedler, P., Brodkorb, S., Fonseca, C., Vaz, F., Zanow, F., & Haueisen, J. (2010). Novel TiN-based dry EEG electrodes: Influence of electrode shape and number on contact impedance and signal quality. In P. D. Bamidis & N. Pallikarakis (Eds.), *XII Mediterranean Conference on Medical and Biological Engineering and Computing 2010* (Vol. 29, pp. 418–421). Berlin, Heidelberg: Springer Berlin Heidelberg. Retrieved from http://www.springerlink.com/content/w8443m31gk8322r0/

g.tec. (n.d.). g.tec medical engineering. Retrieved February 27, 2012, from http://www.gtec.at/Products/Electrodes-and-Sensors/g.SAHARA-Specs-Features

Gargiulo, G., Calvo, R. A., Bifulco, P., Cesarelli, M., Jin, C., Mohamed, A., & van Schaik, A. (2010). A new EEG recording system for passive dry electrodes. *Clinical Neurophysiology*, *121*(5), 686–693. doi:10.1016/j.clinph.2009.12.025

Gersten, R., & Dimino, J. (2001). The Realities of Translating Research into Classroom Practice. *Learning Disabilities Research & Practice*, *16*(2), 120–130. doi:10.1111/0938-8982.00013

Gevins, A., Long, H., Sam-Vargas, I., & Smith, M. (1995). *Patent US5724987 - Neurocognitive adaptive computer-aided training method and system*. Retrieved from http://www.google.com/patents/US5724987

Hsu, W.-Y., Lin, C.-H., Hsu, H.-J., Chen, P.-H., & Chen, I.-R. (2012). Wavelet-based envelope features with automatic EOG artifact removal: Application to single-trial EEG data. *Expert Systems with Applications*, *39*(3), 2743–2749. doi:10.1016/j.eswa.2011.08.132

Khatib, F., Cooper, S., Tyka, M. D., Xu, K., Makedon, I., Popović, Z., Baker, D., et al. (2011). Algorithm discovery by protein folding game players. *Proceedings of the National Academy of Sciences*, *108*(47), 18949–18953. doi:10.1073/pnas.1115898108

Klados, M. A., Papadelis, C., Braun, C., & Bamidis, P. D. (2011). REG-ICA: A hybrid methodology combining Blind Source Separation and regression techniques for the rejection of ocular artifacts. *Biomedical Signal Processing and Control*, *6*(3), 291–300. doi:10.1016/j.bspc.2011.02.001

Liao, L.-D., Chen, C.-Y., Wang, I.-J., Chen, S.-F., Li, S.-Y., Chen, B.-W., Chang, J.-Y., et al. (2012). Gaming control using a wearable and wireless EEG-based brain-computer interface device with novel dry foam-based sensors. *Journal of Neuroengineering and Rehabilitation*, *9*, 5. doi:10.1186/1743-0003-9-5

Liao, L.-D., Wang, I.-J., Chen, S.-F., Chang, J.-Y., & Lin, C.-T. (2011). Design, Fabrication and Experimental Validation of a Novel Dry-Contact Sensor for Measuring Electroencephalography Signals without Skin Preparation. *Sensors (Basel, Switzerland)*, *11*(6), 5819–5834. doi:10.3390/s110605819

Luu, P., Tucker, D. M., & Stripling, R. (2007). Neural mechanisms for learning actions in context. *Brain Research*, *1179*, 89–105. doi:10.1016/j.brainres.2007.03.092

Matsuo, F., Peters, J. F., & Reilly, E. L. (1975). Electrical phenomena associated with movements of the eyelid. *Electroencephalography and Clinical Neurophysiology*, *38*(5), 507–511. doi:10.1016/0013-4694(75)90191-1

Mayers, R. S. (2010). On Teacher Quality. *The Educational Forum*, *74*(3), 272–273. doi:10.1080/00131725.2010.483919

McNamara, D. S. (2006). Bringing Cognitive Science into Education, and Back Again: The Value of Interdisciplinary Research. *Cognitive Science*, *30*(4), 605–608. doi:10.1207/s15516709cog0000_77

Priyadharsini, S. S., & Rajan, S. E. (2012). An efficient soft-computing technique for extraction of EEG signal from tainted EEG signal. *Applied Soft Computing*, *12*(3), 1131–1137. doi:10.1016/j.asoc.2011.11.010

Samuels, B. M. (2009). Can the Differences Between Education and Neuroscience be Overcome by Mind, Brain, and Education? *Mind, Brain, and Education*, *3*(1), 45–55. doi:10.1111/j.1751-228X.2008.01052.x

Savage, N. (2012). Gaining wisdom from crowds. *Commun. ACM*, *55*(3), 13–15. doi:10.1145/2093548.2093553

SFN Report. (2009). *Neuroscience Research in Education Summit: The Promise of Interdisciplinary Partnerships Between Brain Sciences and Education* (pp. 1–7).

University of California, Irvine: Society for Neuroscience. Retrieved from http://www.sfn.org/index.aspx?pagename=NeuroEd_Summit

Smith, M. E., McEvoy, L. K., & Gevins, A. (1999). Neurophysiological indices of strategy development and skill acquisition. *Cognitive Brain Research*, *7*(3), 389–404. doi:10.1016/S0926-6410(98)00043-3

Stripling, R., Becker, W., & Cohn, J. (2005). A Low-Cost , " Wireless ", Portable , Physiological Monitoring System to Collect Physiological Measures During Operational Team Tasks, *000*.

Szűcs, D., & Goswami, U. (2007). Educational Neuroscience: Defining a New Discipline for the Study of Mental Representations. *Mind, Brain, and Education*, *1*(3), 114–127. doi:10.1111/j.1751-228X.2007.00012.x

Tzyy-Ping Jung, Makeig, S., Stensmo, M., & Sejnowski, T. J. (1997). Estimating alertness from the EEG power spectrum. *IEEE Transactions on Biomedical Engineering*, *44*(1), 60–69. doi:10.1109/10.553713

Whitton, J. L., Lue, F., & Moldofsky, H. (1978). A spectral method for removing eye movement artifacts from the EEG. *Electroencephalography and Clinical Neurophysiology*, *44*(6), 735–741. doi:10.1016/0013-4694(78)90208-0

Wilson, G. F., & Fisher, F. (1995). Cognitive task classification based upon topographic EEG data. *Biological Psychology*, *40*(1–2), 239–250. doi:10.1016/0301-0511(95)05102-3

Woestenburg, J. C., Verbaten, M. N., & Slangen, J. L. (1983). The removal of the eye-movement artifact from the EEG by regression analysis in the frequency domain. *Biological Psychology*, *16*(1–2), 127–147. doi:10.1016/0301-0511(83)90059-5

CHAPTER 30

A Neuroergonomic Evaluation of Cognitive Workload Transitions in a Supervisory Control Task Using Transcranial Doppler Sonography

Kelly Satterfield, Tyler Shaw, Raul Ramirez, Erica Kemp

George Mason University
4400 University Drive
Fairfax, VA 22030, USA

ABSTRACT

While automated systems can improve safety and efficiency, these systems can also result in reduced vigilance and reduced situational awareness. As a result of these drawbacks of automation, some research has explored the possibility of adaptive automation as a way to dynamically change function allocation during system operations. Previous research has shown that cerebral blood flow velocity (CBFV), as measured by Transcranial Doppler Sonography (TCD), is sensitive to changes in task demand. Because of these findings, TCD may be a neuroergonomic measure that can be used to assess attentional resource allocation during task performance and is a candidate to add to the battery of measures that can implement adaptive automation. In the current study, participants performed long duration, supervisory control tasks under varying levels of taskload. In one group, enemy threats increased once late in the simulation, and in another group enemy threats increased at two points, once early and once late within the scenario. All participants completed a comparison condition in which there was no increase in the number of enemy threats; they incurred at a steady pace. Performance was assessed by the ability of the operator to protect a no-fly zone over the course of the scenarios. The percentage of red-zone enemy incursions was higher during the transition conditions and cerebral blood flow velocity (CBFV) increased in a similar manner. The results are interpreted in terms of a resource theory of task performance.

Keywords: Transcranial Doppler Sonography, Cerebral Blood Flow Velocity, Resource theory, Mental Workload, Workload Transitions.

INTRODUCTION

Automated systems have become an integral part of operational environments because these systems have been shown to improve safety and efficiency. However, there are drawbacks to the use of automation. Situation awareness is often reduced, there is often a reduction in vigilance, and the use of automation can result in operator skill loss (see Parasuraman & Riley, 1997 for a review). These drawbacks of automation are most salient in the event of a system failure, where operators must transition from periods of very low activity to periods of very high activity. Because of the high cost of such failures, the National Research Council has identified transitions in task demand as a critical concern for the human factors area (Huey & Wickens, 1993). Thus, an understanding as to how operators can cope with taskload transitions while interacting with automation is critical for human-machine reliability and safety.

Physiological measures can provide a way to quantify transitions in mental workload. These measures are an alternative for examining attentional load during dynamic tasks than self-report assessments such as the NASA-TLX (Hart & Staveland, 1988) because of their relatively unobtrusive nature. For example, task performance must be interrupted to administer these scales and can lead to reductions in performance (Moroney, Biers, and Eggemeier, 1995). One candidate for physiological assessment during performance is Transcranial Doppler Sonography (TCD). Evidence for the feasibility of using TCD as a measure of load comes from the vigilance/ CBFV relation. A series of research studies using the TCD procedure in studies of sustained attention have shown that the technique can provide a metabolic index of attentional resource utilization during task performance (e.g. Hitchcock et al., 2003; Warm and Parasuraman, 2007). These studies, which featured CBFV measurements in the right and left middle cerebral arteries, have consistently shown that the vigilance decrement, the decline in signal detections over time that typifies vigilance performance, is paralleled by a decline in CBFV (e.g. Hitchcock et al., 2003; Shaw et al, 2009, Shaw et al., 2012). In addition, the absolute level of CBFV in these studies was directly related to the cognitive and psychophysical demands of the vigilance task, and the overall effects were lateralized to the right cerebral hemisphere.

The resource model derived from the vigilance/CBFV relation has recently been extended to more dynamic tasks that require more complex decisions, as compared to the binary decisions (signal present/absent) that are required in a vigilance task. Shaw et al. (2010) have shown that cerebral blood flow velocity is increased following transient increases in task load in a command and control simulation environment called the Distributed, Dynamic, Decision-making task (DDD). In that study, participants controlled six friendly UAV assets. During each 10-minute scenario, operators were tasked with destroying enemy threats to prevent incursions into a no-fly zone. There were two different levels of task load. In the low taskload condition, enemies presented steadily throughout the 10-min scenario. The high task load condition featured two points in the scenario in which the frequency of enemy presentation was unpredictably increased. Results showed that as the taskload was unexpectedly increased, the CBFV measure also increased.

While CBFV *increased* as task load increased in the Shaw et al (2010) study, one expectation in that study that was not met was that as task load decreased, there would be a subsequent decrease in CBFV. This finding would not appear to support the aforementioned resource model, which posits that an increase in CBFV should result in a subsequent decrease in CBFV, especially when task demand is lowered. Furthermore, because there should be drainage of attentional resources, fewer should be available to allocate to the task. Alternatively, another way to characterize this effect would be with an *effort regulation* model of taskload transitions. In that model, the inability to effectively manage the reduction in task load is a result of the operator failing to match the allocated level of effort to task demands (Hancock & Warm, 1989; Matthews & Desmond, 2002). In other words, the operators in the Shaw et al (2010) study may have overestimated the demand of the task and may have applied more effort than necessary, thereby resulting in elevated levels of CBFV for low task demands. It would appear that the latter model is a better fit for characterizing that data.

The goal of the current study was to test the alternative predictions made by the two aforementioned theories in an experiment that measures CBFV during the performance of a long-duration supervisory control simulation. To accurately test the two theories, the length of the scenarios was extended from a relatively short-duration (10 minutes) to a much longer-duration (30-minutes). Furthermore, the time that occurred *between* transitions was extended. Unlike in the Shaw et al (2010) study, the CBFV data in the current study were also compared to a performance measure, and not merely changes in simulation events. Under the resource model, it would be predicted that as task demand increases *and* decreases, CBFV should increase and decrease in a similar manner. Furthermore, if there are two transitions within a single scenario, the CBFV change to the second transition should be less robust than the first because of a reduction in the availability of information processing resources. Alternatively, the effort regulation model predicts a potential increase in task performance, but a decrease in task demand will either be delayed or will not occur because observers may inappropriately allocate their effort. It was expected that this follow-up study to the Shaw et al. (2010) would be supportive of resource theory.

METHOD

Participants
Twenty undergraduate students (5 male and 15 female) participated in this study. All participants were given credit for their participation. Participants completed two experimental trails presented in random order. Each participant completed a low workload condition and workload transition condition. In the low workload condition, enemies were presented at a rate of 1.5 per minute, on average, with 45 total enemy threats being presented in each 30 minute trial. In the one workload transition condition, enemies were presented at a rate of 1.97 per minute, on average, with 59 total enemies, and the most enemy threats presenting at minute 22. In the two workload transitions condition, enemies were presented a rate of 2.4 per minute, on average, with 72 total enemies, with the most enemy threats presented at minute 7 and minute 22.

DDD Simulation
This experiment used the Dynamic Distributed Decision Making (DDD) 4.0 simulation, developed by Aptima Inc which is a tool used to create distributed, multi-person, and automation based simulations. Participants performed the DDD scenario at a Desktop computer running Windows XP with a 17-inch screen. Each scenario was a 30-minute simulated air operation where participants controlled six friendly UAV fighter assets. Neutral and enemy targets could enter the green zone and move toward the red friendly zone from any direction. Enemies could then attack and destroy friendly assets. Participants were instructed to: 1) defend against enemy targets entering the red zone, 2) destroy enemy targets as quickly as possible, and 3) protect own friendly assets from being destroyed.

Participants first participated in computer based training that involved instructions presented via PowerPoint that explained the DDD simulation and the experiment. Participants were then shown a tutorial that demonstrated how to select and move friendly assets, as well as destroy enemy targets. After the tutorial, participants completed two five minute practice trails, one with a low workload and the second with a higher level of workload. There were no taskload transitions during training. After the practice trails, participants completed two experimental trials.

Prior to the start of the first experimental trial, participants were hooked up to a TCD unit equipped with two 2-MHz ultrasound transducers. The transducers were attached tightly to an adjustable plastic headset and were used to take hemovelocity measurements in cm/sec of the right and left middle cerebral arteries. The transducers were placed dorsal and immediately proximal to the zygomatic arch along the temporal bone. In order to retain transmission of the ultrasound signal, a small amount of ultrasound gel was placed on the transducers. The participant remained connected to the TCD unit for the entire duration of the experiment, approximately 1 hour and 30 minutes.

RESULTS

Performance

The percentage of enemy incursions into the red-zone was used to characterize operator performance, thus, lower numbers indicate better performance. To specifically test the hypothesis that performance would vary directly with the transition conditions, data for each 30 minute scenario were collapsed across 5 phases: Pre-transition (minutes 1-5), Transition 1 (minutes 5-9), Post-transition 1 (minutes 10-18), Transition 2 (minutes 19-23) and post Transition 2 (minutes 24-30). Each of the 3 conditions were analyzed separately by means of an analysis of variance (ANOVA) conducted on the percentage of enemies allowed into the red-zone. The analysis of the single workload condition revealed a significant main effect for Period, $F(2.3, 20.71) = 5.88$, $p < .05$. In this and all subsequent ANOVAs, Box's epsilon was used to correct for violations of the sphericity assumption. Post-hoc tests revealed that percentage of enemies allowed into the red-zone did not differ between Pre-Transition ($M = 6.7\%$), Transition 1 ($M = 8.0\%$), and Post-Transition 1 ($M = 4.3\%$). There was, however, a significant decrease in performance from Post-Transition 1 ($M = 4.3\%$) to Transition 2 ($M = 19.2\%$), and performance significantly increased from Transition 2 to Post-Transition 2 ($M = 2.2\%$). The ANOVA of the two workload transition condition also revealed a significant main effect for the percentage of enemies allowed into the red-zone for Period, $F(1.28, 11.52) = 14.63$, $p < .05$. Post-hoc tests revealed that the percentage of enemies allowed into the red-zone was significantly increased from Pre-Transition 1 ($M = 4.6\%$) to Transition 1($M = 42.1\%$) and from Post-Transition 1 ($M = 7.1\%$) to Transition 2 ($M = 15.0\%$). Additionally, post-hoc tests revealed that the percentage of enemies allowed into the red-zone was significantly reduced between Transition 1 to Post-transition 1, and again from Transition 2 to Post than and Post-Transition 2 ($M = 6.7\%$). The ANOVA for the low workload condition on the percentage of enemies allowed into the red-zone did not reveal any significant changes for Phase, $p < .05$. The data of red zone performance for the single, dual, and no transition conditions can be viewed graphically in Figures 1, 2, and 3 respectively.

Blood Flow Velocity

To account for the wide range of hemovelocity scores present in the population, the scores were expressed as a proportion of the last 60 seconds of a five minute resting baseline (Warm & Parasuraman, 2007). Similarly to performance data, CBFV data were collapsed across the five phases of Pre-Transition, Transition 1, Post-Transition 1, Transition 2, and Post-Transition 2. A 2 (Hemisphere) x 2 (Condition) x 5 (Period) ANOVA performed on the CBFV scores of the single transition condition revealed a significant main effect for phase,

$F(2.93, 52.65) = 0.00$, $p < .05$. Post-hoc tests revealed that CBFV did not differ between Pre-Transition (M = .99) Post-Transition 1 ($M = .98$), and Post-Transition 2 ($M = .98$). There was, however, a significant increase in CBFV from Pre-Transition ($M = .99$) as well as from Post-Transition 1 ($M = .98$) to Transition 2 ($M = 1.00$). The ANOVA performed on the dual taskload condition revealed a significant main effect for phase, $F(2.47, 19.79)$, $= 5.53$, $p < .05$. Post hoc tests revealed that there was an increase in CBFV from Pre-transition ($M = .99$) to Transition 1 ($M = 1.03$), a significant decrease from Transition 1 ($M = 1.03$) to Post-Transition 1 ($M = .97$). Finally, there was a significant decline from Transition 2 ($M = 1.00$) to Post-Transition 2 ($M = .97$). The ANOVA performed on the low workload condition revealed no significant effects or interactions, $p < .05$ The data of CBFV for the single, dual, and no transition conditions can be viewed graphically in Figures 1, 2, and 3 respectively.

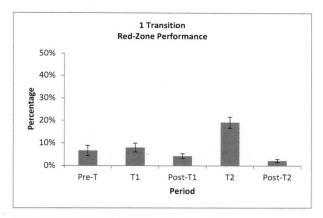

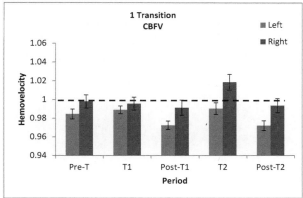

Figure 1. Performance and CBFV data for the one transition condition. Top panel: Performance scores as the percentage of red-zone enemy incursions plotted as a function of five periods. Bottom Panel: Hemovelocity scores relative to baseline plotted as a function of five periods. The dashed lines represent baseline. Error bars are standard error.

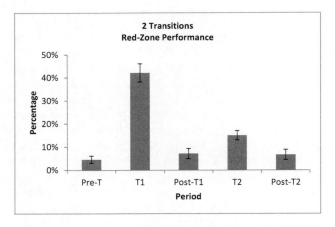

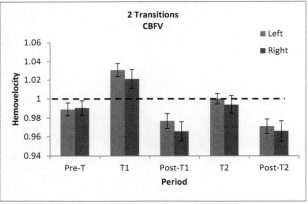

Figure 2. Performance and CBFV data for the two transitions condition. Top panel: Performance scores as the percentage of red-zone enemy incursions plotted as a function of five periods. Bottom Panel: Hemovelocity scores relative to baseline plotted as a function of five periods. The dashed lines present baseline. Error bars are standard error.

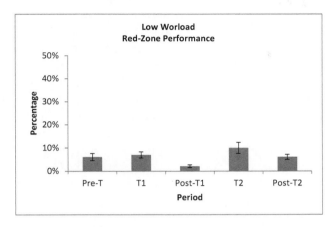

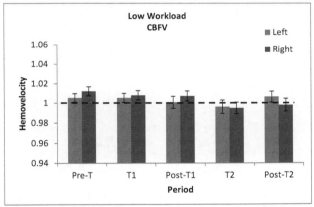

Figure 3. Performance and CBFV data for the low workload condition. Top panel: Performance scores as the percentage of red-zone enemy incursions plotted as a function of five periods. Bottom Panel: Hemovelocity scores relative to baseline plotted as a function of five periods. The dashed line represent baseline. Error bars are standard error.

DISCUSSION

This study was designed to determine the degree to which performance in dealing with unpredictable changes in taskload is paralleled by changes in CBFV. The CBFV data is consistent with previous findings that the CBFV measure can be used as an index of information processing resources in vigilance (Warm & Parasuraman, 2007) and in supervisory control settings (Shaw et al., 2010). Results showed that as task demands increased, operators expended more cognitive resources as reflected in an increase in CBFV. Similarly, when task demands decreased, CBFV decreased in a similar manner suggesting that operators are not expending as many resources. Lastly, when task demand remained constant, CBFV remained constant as well.

These results support an attentional resources view of task performance. The effort-regulation theory of performance would have predicted that workload should remain high after the initial increase in task demand because observers are failing to appropriately match the level of effort expenditure to the demands of the task. This maintenance of high workload should be reflected in elevated levels of CBFV, even after subsequent decrease in taskload. The data show that there is a clear parallel between the performance and CBFV measure, contrary to the predictions made by the effort regulation and in support of a resource model of performance.

This study provides a demonstration that the CBFV measure is an index of cognitive resources and TCD can be used to detect moment to moment workload variations. These findings have important implications in such areas as sustained monitoring of operational tasks in air traffic control and airport baggage screening. Such tasks often involve fluctuations in task demand and often involve large costs associated with errors. The CBFV measure offers a way to assess the cognitive resources involved in high workload demand particularly in fields where errors can result in disastrous consequences.

REFERENCES

Hancock, P. A., & Warm, J. S. (1989). A dynamic model of stress and sustained attention. *Human Factors, 31*, 519-537.

Hart, S., & Staveland, L. (1988). Development of NASA-TLX (Task Load Index): Results of empirical and theoretical research. In P. Hancock & N. Meshkati (Eds.), Human mental workload (pp. 139-183). Amsterdam: North Holland.

Hitchcock, E.M., Warm, J.S., Matthews, G., Dember, W.N., Shear, P.K., Tripp, L.D., Mayleban, D.W., & Parasuraman, R. (2003). Automation cueing modulates cerebral blood flow and vigilance in a simulated air traffic control task. *Theoretical Issues in Ergonomics Science, 4*, 89-112.

Huey, B.M., & Wickens, C.D. (1993). *Workload transition: Implications for individual and team performance.* Washington, DC: National Academy Press.

Matthews, G., & Desmond, PA (2002). Task-induced fatigue states and simulated driving performance. *Quarterly Journal of Experimental Psychology, 55*, 659-686.

Moroney, W. F., Biers, D. W., & Eggemeier, F. T. (1995). Some measurement and methodological considerations in the application of subjective workload measurement techniques. *International Journal of Aviation Psychology, 5*, 87-106.

Parasuraman, R., & Riley, V. (1997). Humans and automation: Use, misuse, disuse, abuse. *Human Factors, 39*, 230–253.

Shaw, T.H., Parasuraman, R., Sikdar, Siddhartha, & Warm, J.S. (2009). Knowledge of Results and Signal Salience Modify Vigilance Performance and Cerebral Hemovelocity. *Proceedings of the Human Factors and Ergonomics Society, USA, 53,* 1062-1065.

Shaw, T. H., Warm, J. S., Finomore, V. S., Tripp, L., Matthews, G., Weiler, E., & Parasuraman, R. (2009). Effects of sensory modality on cerebral blood flow velocity during vigilance. *Neuroscience Letters, 461*, 207-211.

Shaw, T.H., Guagliardo, L., de Visser, E., Parasuraman, R. (2010). Using Transcranial Doppler sonsography to measure cognitive load in a command and control task. *Proceedings of the Human Factors and Ergonomics Society, 54,* 249-253.

Shaw, T.H., Finomore, V.S., Warm, J.S., Matthews, G. (2012). Effects of regular or irregular event schedules on Cerebral Hemovelocity during a sustained attention task. *Journal of Clinical and Experimental Neuropsychology, 34,* 57-66.

Warm, J.S., & Parasuraman, R. (2007). Cerebral hemovelocity and vigilance. In R. Parasuraman & A.M. Rizzo (Eds.), *Neuroergonomics: the brain at work.* Cambridge, MA: MIT Press.

CHAPTER 31

From Biomedical Research to End User BCI Applications through Improved Usability of BCI++ v3.0

Paolo Perego[a], Emanuele Gruppioni[b], Federico Motta[a], Gennaro Verni[b], Giuseppe Andreoni[a]

[a]Indaco Department, Politecnico di Milano, Milano, Italy
[b]INAIL Centro Protesi, Vigorso di Budrio (BO), Italy
paolo.perego@polimi.it

ABSTRACT

Last decade showed a high evolution in Brain-Computer Interface (BCI) system development, starting from the algorithms for real-time processing (thus increasing the bit-rate [Ryan 2010]) up to new frameworks for the creation of this BCI systems.

On the market we can find many frameworks (Brunner et al 2010), some of them are recent projects (openVibe, Tobi) and can be used on different computer architectures, while other ones like BCI2000 are now widely used in many research centers around the world. The research in this field focused on evaluating accuracy of a BCI system in some daily life applications, without considering the importance of the Human Computer Interaction (HCI). For this reason we can find complex and well-performing Brain-Computer Interface systems with a high bit-rate but which cannot be used in everyday life because they are developed to work in controlled environments and on specific, technologically-difficult platforms (i.e. definitely not easy-to use except by skilled technicians). In this paper we present our new BCI++ v3.0 framework, aiming at giving to researchers and developers a new tool to create BCI system focusing on the creation of GUI which are simple to use, intuitive and at the same time well-designed, and functioning also for applications with few commands and speed of use (bitrate). Our framework, developed with Microsoft© XNA Game Studio, provides some useful tools to create an immersive environment which could be used not only for BCI purposes, but also for biomedical signal analysis and biofeedback. Some prebuild GUI (like P300 oddball protocol, SSVEP protocol (Parini et al 2009) can help the developer design the system.

1 INTRODUCTION

A Brain Computer Interface (BCI) is a man-machine-interface which allows to establish a new communication channel between the brain and the computer (Wolpaw *et al.* 2002). It is dedicated to control of computers or external devices with real-time monitoring, processing and classification of the only brain activity. This technology could be used in order to restore or to support lost body functions and it could potentially represent a substantial improvement to the quality of life of severe disabled people where also basic functions (verbal communication or motor activities) are compromised. In recent years many different research centers built various typologies and architectures of BCI systems, focusing especially on the performance, and demonstrating the ability to use the BCI into some daily life applications. Nevertheless all these systems still showed their usability limits being far from being used out of a controlled environment such as a research laboratory; this is mainly due to the fact that high specialization is required by the operators in using the different setup tools. Berger et al 2007 predicted a mass market in a short time, but in the last ten years only one BCI system, the IntediX by Gtec (www.intendix.com), has been released in the market.

The BCI++ framework was born in 2008 (Maggi *et al.* 2008) with the primary goal to provide researchers and developers a new and flexible tool to create BCI system focusing on the creation of GUI which are simple to use, intuitive and at the same time well-designed for application with few commands and speed of use (bitrate). For these reasons the user interface development platform of BCI++ was based on an open-source graphics engine (Irrlicht) (Perego *et al.* 2009) which, with the use of software like IrrEdit, allowed the visual development of the interface. Moreover al the framework was developed using the most widespread programming language, the C/C++, in order to simplify even further the development of systems and to allow changes even at low level (graphics, communication protocol...). With this configuration BCI++ had a big limitation: the framework used library only for Windows© and the software IrrEdit was not easy to use; moreover the plug-in structure of the system didn't allow an easy integration with other tools (eg. Augmented Reality, Kinect).

For these reasons we extend the usability and the performance of BCI++ using a new graphics engine, or rather a framework, which in the last two years has been widely spread: Microsoft XNA game studio.

2   THE SYSTEM

Microsoft XNA Game Studio can be successfully used for BCI purposes because it includes a Graphics Engine, a Sound Engine and high and low level GUI programming. XNA allows also to create multi-platform user Interface which can be used on PC, windows phone 7 devices and XBOX360. The release of Kinect SDK and other library allow to insert easily microphone and motion and gesture capture supports into BCI systems.

This work aimed at integrate the XNA framework into the BCI++ user interface and redesign the hardware interface acquisition software user interface for a simpler using. The BCI++ v3.0 has the same structure of the previous version (Perego et al 2009): AEnima, the user interface software, has been replaced by XAEnima which include XNA while HIM, the hardware interface module, has undergone graphic restyling.

The two software can run both on the same PC or on two separated different ones, and they communicate via TCP/IP socket connection.

Unlike previous versions which were very closed, the use of new software libraries allows the easy integration of new components within the system, such as haptic interfaces (like the Wii balance board) or other input systems.

3   THE ACQUISITION SOFTWARE

The Acquisition software is the main core of the BCI system. It is an open source GPL software designed in 2008 in order to provide a solid structure for the acquisition, storage , processing (both via C/C++ and Matlab Mathworks© algorithms) and visualization of different kind of signal, both biomedical and that non-medical (acceleration). HIM v8.0 (this is the name of the software) was initially designed only for BCI, for this reason we integrate the communication with different kind of EEG amplifier; at now it supports several kind of instruments, not only EEG but also for various kind of signals. The compatible devices are:

- BrainProducts VisionRecorder
- BrainProcuts Vamp USB
- G.tec g.Mobilab
- G.tec g.USBamp
- Compumedics Neuroscan
- Braintronics BrainBox 1xxx
- SXT Telemed ProtheoII
- SXT Telemed Phdra
- Emotiv (soon)

It supports also some open-source hardware like Arduino ECG/EEG which allow to acquire bio-signals with low cost systems. The software has an help and a wizard to help developer to add other devices not present into the list.

For prototyping purpose, we include also virtual signal generator which allow to play pre-recorded file or generate sin waveform. HIM can communicate with the supported devices both via Bluetooth® and TCP/IP.

Thanks to the supported devices, during the last years HIM has been widely spread and it is used in several research laboratories all over the world. Thanks to this, and to the information recorded during the thousands of acquisitions made with this tool, it was possible to create a new graphical interface to improve the usability of the software. Figure 1 show the new look of the GUI for the acquisition software.

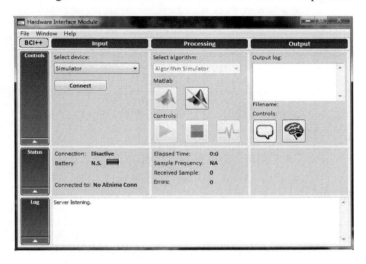

Figure 1: The new look of HIM v8.0; the matrix structure allow to easily understand the software operation making it easier to use by unskilled subjects.

As shown, the new graphical interface was structured as a "Matrix". We have three rows and three columns. The rows group the interface elements by action: controls, status and log; the columns group the elements by progress: input, processing, output.

The first row, the controls, includes all the element necessary to act on the system, like the start and stop acquisition button, the element for the device connection and the selection of algorithm. The second row, the status, lists all the status variable; the connection type, the battery level (if present), and all the information regarding the acquisition like the sample frequency, the errors and the elapsed time. The last row, the log, reports all the action occurred during the acquisition (connection and disconnection, low battery…).

Instead the column allow to quickly identify the progress of the signals, so the user can find the input with all the elements regarding the acquisition device; the processing includes the algorithms and all the elements which allow to control the acquisition; finally the output includes all the elements that can show the output (processed signals or feedback) both in graphic or text (output log) style.

Obviously all these elements and controls are not always necessary, some can be used only by advanced user to setup the system but are useless for end users. For this reason we have implemented controls on the GUI (the left grey button in Figure 1) which allow to hide part of the graphical interface simplifying the interaction. Figure 2 show the reduced graphical interface with only the status information visible.

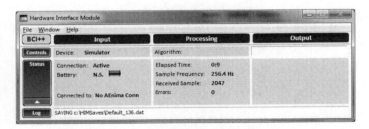

Figure 2: HIM software with reduced and simplified graphical interface.

4 THE USER INTERFACE SOFTWARE

The user interface is a software with a set of libraries which allow simplifying the implementation of new graphical interface for BCI operating protocols. XAEnima has a software core, developed by means of XNA (for this reason the name XAenima), which includes all the functions for the communication (via TCP/IP [Perego et al 2009]) with the BCI++ acquisition module.

The XAEnima is an independent application written in C# language (for the simple connection with XNA). The two modules, HIM and AEnima, are connected via TCP/IP and are thought to run on two different computer, as well is possible to run on the same one. Furthermore XAEnima is an independent software and as such it can be used separately for example when the signal processing is implemented directly on the acquisition device; thereby the acquisition software is not necessary.

The communication with the acquisition software was implemented by means of a simple protocol in which any message was composed by a kind that identifies the action to be done (start acquisition, stop acquisition, classification, rest…) followed by an array containing the values that will be used for that action. This protocol allow both the communication between the two software, but can be used also for control any device that can be connected to the system.

XNA makes easier to implement graphical user interfaces thanks to the many existing tools and to the presence of online countless resources and examples.

Figure 3 shows the quality of the new framework; in the figure you can view the p300 oddball keyboard created by means of the BCI++ with XNA Game studio.

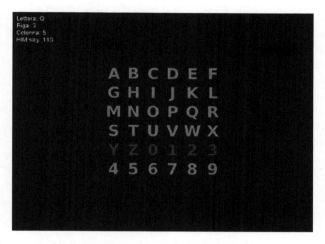

Figure 3: p300 oddball keyboard with XAEnima

XAEnima includes also some precompiled tools which can be useful for the development of BCI; we have included some different visualization of biofeedback structured into different forms:

- Bars;
- Circle;
- Sound;
- Other (through external connection via USB/Bluetooth).

XAEnima presents also some functions for the stimuli used in P300 and SSVEP (Wolpaw et al 2002) BCI protocols: we implements Oddball keyboard for P300 and checkerboard inversion or simple square SSVEP on-screen stimuli. It also includes a library for SSVEP external stimulator (via USB) whose schematics is available online as open-source.

Thanks to the XNA Game Studio capability, XAEnima can be compiled (it is released under GNU GPL license) and used on PC, XBOX360, Windows 7 Phone or Tablets.

6   IMPLEMENTATION EXAMPLES

The new BCI++ v3.0 framework has been used to implement two SSVEP protocol based BCI systems with two different purposes: Home Automation Control and Rehabilitation.

The first one uses both external and on-screen stimulator to elicit the SSVEP pattern in order to control, by means of a menu, a complete home automation system (MyHome - Biticino). The home automation system has been connected via TCP/IP with a webserver and can control different kind of actuator in order to allow to the disabled persons to control virtually any kind of device in his home. Thanks to the BTicino webserver the home automation can work in two ways: the disabled can control the home improving his style and quality of life, while the relatives, using a protected internet connection, can monitor whether the disable person is comfortable. The system is in the test phase and will be replicated at the INAIL Prosthesis Center in Vigorso di Budrio (BO, Italy) with the aim to test the entire system (software and external component) in real life applications.

The second one was developed in collaboration with the Informatics Department of the Politecnico di Milano and consists of a simple maze videogame: the subject, using first BCI and then the Wii Balance Board, has to complete the maze in the least time as possible. This application was tested with SSVEP protocol in order to simplify the testing phase (SSVEP paradigm requires a short training time), but it has been designed to be used with motor imagery BCI in order to accomplish a complete rehabilitation exercise. Figure 4 shows two screen of the rehabilitation game.

Figure 4: Example of BCI++ v3.0 GUI. XNA games studio features allow at creating immersive graphic interface, from high quality menu to simple 2d game.

5 CONCLUSION

The new BCI++ v3.0 framework has been validated by setting up the two BCI systems. However the usability of the graphics interface was not evaluated yet. We are designing the multi-factorial protocol in order to evaluate the usability of the GUI and the stimulation methods by means of an eye tracking system.

ACKNOWLEDGMENTS

This work was partially supported by the INAIL Centro Protesi (Budrio, Bologna, Italy). The authors thank eng. C. Motta and J. Yuan for their precious help in the development and the debugging of the rehabilitation game, eng. Enrico Valtolina and eng. Massimo Perego of BTicino spa (Italy) for having generously provided the MyHome system and engineers L. Maggi and S. Parini for all the support in work.

REFERENCES

Brunner C, Mellinger J, Schalk G, Renard Y, Lecuyer L, Breitwieser C, Mueller-Putz GR, Kothe CA, Makeig S S, Perego P, Andreoni G, Kanoh S, Susila IP, Bianchi L, Venthur B, Blankertz B. 2011, "BCI Software Platforms." in *The Human in Brain-Computer Interfaces and the Brain in Human-Computer Interactions*, ed. Tan D & Nijholt A, Eds. (B+H)CI edn, Springer Verlag, in press.

BTicino "Myhome biticino." [Online]. Available: http://www.myhome-bticino.it

Maggi L , Parini S, Perego P, Andreoni G. 2008, "BCI++: an object-oriented BCI Prototyping Framework", Proceeding of 4th International Brain-Computer Interface Workshop 2008 - Graz , September 19-20.

Perego P, Maggi L, Parini S, Andreoni G. 2009, "BCI++: A New Framework for Brain Computer Interface Application", Proceeding of 18th International Conference on Software Engineering and Data Engineering (SEDE-2009), ed. ISCA, ISCA, Las Vegas (USA), 22-24 June.

Ryan D B 2010, "Predictive Spelling With a P300-Based Brain–Computer Interface: Increasing the Rate of Communication", *International Journal of Human-Computer Interaction,* 27:69-.

Wolpaw JR, Birbaumer N, McFarland DJ, Pfurtscheller G, Vaughan TM. 2002, "Brain-computer interfaces for communication and control", *Clinical neurophysiology : official journal of the International Federation of Clinical Neurophysiology,* 113:767-791.

CHAPTER 32

An ERP Study on Instrument Form Cognition

Ying Jiang[1], Jun Hong[1], Xiaoling Li[1], Wei Wang[2]

[1]Xi'an Jiaotong University, [2]Missile Institute of Air Force Engineering University
Shaanxi, PRC
samanthajiang@gmail.com

ABSTRACT

In order to explore the influence of different types of instruments and color combinations on brain activity, with speedometer as the example, four kinds of dial shapes and four design variables were adapted and studied. Thirty healthy graduate students participated in the study. Participants were required to react by pressing specific keys, respectively, when a pointer pointed to a safe area versus a warning area of a dial; reaction time (RT) and accuracy were recorded. Within each dial shape, combinations with shorter RT and higher accuracy (class-G) and combinations with longer RT and lower accuracy (class-B) were selected and subsequently used as target stimuli for which event-related-potentials (ERPs) were recorded. Through t-tests and ANOVA, similar trends in class-G on P300 amplitude at Fz, Cz, Pz and Oz illustrated that brain activity induced by different dial shapes is similar, and dials in class-B were found to need more attention to read. Furthermore, frontal and parietal alpha band power appears to be significantly higher under class-B. It is worth noting that changed location of warning area reflected the difference of band power.

Keywords: event-related-potential (ERP); Instrument; Cognition; P300 amplitude; band power

1 INTRODUCTION

With the development of computer technology, a great change has taken place in instrument forms. The visual display terminal (VDT) has become the primary form of contemporary instruments. Meanwhile, increases in information displayed and display speed could lead to an increase in operator error rate. As a result, the impact of human cognitive and perceptual activities on display performance with VDT is receiving increased attention from researchers.

There has been considerable research on instrument cognition and the impact of instrument characteristics on human reactions (Zhang & Zhuang, 2009; Zhang et al, 2009b; Zhuang & Wang 2003). VDT display color is an important factor for determining how good observers' visual performance is (Garcia & Caldera 1996; Shieh & Lin 2000; Zhang & Zhuang 2009). Color, shape, or pattern may all impact human cognition. For example, the perceived shape of an object is often altered following adaptation to an object of slightly different shape (Anderson et al 2007; Bell et al 2009; Bell & Kingdom 2009; Gheorghiu et al 2010).

Traditionally, methods to study instrument cognition mainly focus on a combination of subjective and objective methods (Senol et al, 2009), with more emphasis generally placed on qualitative approaches (Jiao et al 2007; Wanyan et al 2009). Recently, studies on instrument display design have started from cognitive characteristics, and evaluated quantitative measures of reaction time and accuracy as indicators of human performance, as well as subjective evaluations (Zeng & Zhuang 2006; Zeng et al 2007; Zhang et al 2009a; Zhang et al 2009b; Zhuang & Wang 2003). Mental workload, defined as the difference between the available and required capacity of the human information-processing system to perform a task at any given time, has been considered for the identification of target color, shape, and position under different task demands (Zeng & Zhuang 2006; (Kang et al 2008).

Instrument cognition is a part of human cognition, cognition through vision influences brain activities greatly (Du et al 2009). During instrument cognition, visual searching, attention, and conscious processing may be used, which means during cognitive process a series of mental activity can be produced. Mental activity has the oriented ability of preventing useless information from accessing the consciousness (Wei & Yan 2008). Therefore, changes in mental activity embody differences in information processing mode during instrument cognition processing. Research on mental activity analysis during instrument cognition is lacking.

The main purpose of this study was to explore differences on brain function with instrument cognition under different shape and color combinations. Common circular and linear dials were adapted, with different color combinations. Round and semicircular dials were involved in circular dials; horizontal and vertical dials belonged to linear ones. The warning area and safe area of the dials were marked with different colors. Two tasks were performed in the study. In the first task, participants were required to react by pressing specific keys, respectively, when the pointer pointed to a safe area or warning area, meanwhile, reaction time (RT) and accuracy were collected. Through statistical analysis on the data, good color combinations with shorter RT and higher accuracy and bad color combinations with longer RT and lower accuracy in each dial shape, named class-G and class-B respectively, were selected and used as target stimuli subsequently in the second task. A typical P300 experimental paradigm was adapted (Ma & Gao, 2008) to collect event-related-potentials (ERPs) in order to explore the brain activity induced by instrument cognition. Through observing P300 frequency components and regional distributions of brain activity under different shape dials, we sought to study the attention characteristics associated with instrument cognition.

2 COGNITION AND JUDGEMENT EXPERIMENT

2.1 Methods

2.1.1 Participants

Thirty healthy graduate students (18 males, 12 females), between 20 and 33 (M=23.6, S.D.=2.88) years of age, participated in the study. All participants had normal or corrected to normal vision. They were all right-handed and without any eye diseases. Participants were paid for their participation.

2.1.2 Materials

We adapted four kinds of dial shapes (Tilley & Associations 2001), such as round dial (RD), semicircular dial (SD), horizontal dial (HD), and vertical dial (VD), as well as four design variables (pointer colors, color for safe area scale, color for warning area scale and pointer addressing) for the experimental design (see Table 1). The radius of RDs and SDs was the same as the length of HDs and VDs. In order to better display on a 23-inch LCD computer monitor, all dials were displayed in 2x size except for the VD. Vertical dials were displayed in 1.4x size. All dials were displayed with black background because it can generate better working performance (Hou et al 2008).

Table 1 Levels for the factors

Levels	Pointer Color X1	Safe area color X2	Warning area color X3	Pointer Addressing X4
1	White (255,255,255)	White (255,255,255)	Red (255,0,0)	Safe area
2	Red (255,0,0)	Green (144,238,144)	Yellow (255,255,0)	Warning area

*The numbers in brackets was the value for the contents of Red, Green and Blue.

The experiment used an orthogonal experimental design method (Yuan et al, 2007). Thirty-two dial pictures were represented as stimuli randomly. E-Prime v 2.0 was used to present instructions and a fixation screen to participants during the experiment on a 23 LCD computer monitor, which was 70 cm away from the participant, the angle for an average height participant was 15 degrees.

2.1.3. Procedure

The experiment was divided into four trials according to dial shapes (See Figure 1). When the pointer of the dials pointed to warning areas the participants need to press the left mouse button as quickly and as accurately as possible; if the dial pointed to the safe area, the right mouse button need to be pressed. The choice buttons were alternate in each trial, for example, in trial the 1 left mouse button was used for the warning area, in trial 2 the right mouse button for was used for it.

During the experiment, the stimuli remained on the screen until the participants made a reaction. If participants did not react in 5 minutes after stimulus onset, the next stimulus was presented. Before the appearance of the dial stimuli, a white plus sign was presented in the center of the screen, in order to reduce the short-term memory of the previous display (See Figure 1). Reaction time (RT) and action accuracy were recorded simultaneously. RT was assessed by recording the time delay from the presentation of each stimulus (stimulus onset) to the time of reaction.

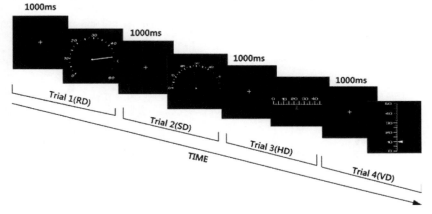

Figure 1 Cognition and judgment task on different shape dials. A complete trial includes four separate trials: four shapes trials.

2.2 Results and analysis

The main analyses were conducted on the average RT and misses (see Table 2) by t-tests and a single factor ANOVA.

The results showed that the difference between RD and SD on RTs was significant $t(15)=2.782$, $p<0.05$; the difference between RD and VD on RTs was significant $t(15)=3.185$, $p<0.001$; the differences among RD and others on Misses were significant, with SD $t(15)=2.787$, $p<0.05$, with HD $t(15)=-3.136$, $p<0.01$, with VD $t(15)=-3.627$, $p<0.01$. The difference among the different shapes on RT and misses was significant $F(3)=3.479$, $p<0.05$, $F(3)=6.692$, $p=0.001$. The above means that the response on RD was significantly different compared with other shapes. The area of the dials influenced RT and the misses directly.

Table 2 Mean RT and the misses under the different shapes

Item	Round Dial	Semicircular Dial	Horizontal Dial	Vertical Dial
RT (ms)	835 (75.2)	778(45.4)	793(88.4)	759(71.8)
Misses (%)	15.2(8.7)	8.1(4.0)	7.3(5.9)	6.7(5.2)

*Misses is the rate of wrong action out the whole action.

From Table 2, mean RTs were 750~850ms, misses did not exceed 20%. In order to examine subtle differences between the different colors under dial shapes, 750ms and 900ms in RT, 5% and 20% in Misses were selected as the boundaries respectively. If RT >900ms, we considered that it was a slower reaction; if RT<750ms, it was treated as a faster reaction. Figure 2 shows the influence of the different colors on RT and misses. The red warning area can evidently promote RT; while yellow slows down reaction time. The white safe area can reduce reaction misses, while the green safe area and yellow warning area increase misses. In a word, color for warning area scale evidently affects RT, while the effect from pointer color and color for safe area scale is not obvious; the effect on the misses from scale color is significant, while the effect from pointer color is not significant.

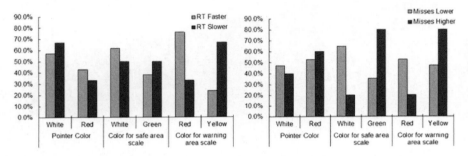

Figure 2 Comparison of difference on RT and misses under different color combinations. The percentage stands for the influence degree of the color on indicators (RT and Misses).

2.3 Discussion

When warning area scale color is red, if the pointer color is red or safe area scale color is white, the RT will be much shorter; but if the pointer color is white or safe area scale is green, the RT has the trend to be longer, while the effect on accuracy is not significant (Table 3). Compared with red, when warning area scale color is yellow, the RT will be longer and the accuracy will be declined. But if yellow warning area is coupled with a white pointer, the situation will be improved; if it works with green safe area, the situation will become worse. Besides, when pointer color is close to warning area color, RT and accuracy all will be improved. Similar with the circumstance of pointer color, white pointer mainly influences RT and red pointer influences reaction misses. Changes of pointer addressing do not lead to any significant variations.

In a word, if the pointer color was the same as the color of the scale which it was addressing to, the work efficiency will be improved, but the pointer color does not influence the indicators (RT and accuracy) separately. The color for safe area scale can influence the accuracy significantly, meanwhile under the interaction with color for warning area scale; the influence on two indicators is significant. Color for warning area scale has the largest influence on the indicators.

Table 3 Rate of change on indicators under different color combinations

Rate of change	Pointer color	Color for safe area	Color for warning area	Pointer color-Safe area scale color Combination				Safe area scale-warning area scale color Combination			
				White-Red	White-yellow	Red-Red	Red-Yellow	White-Red	White-yellow	Green-Red	Green-Yellow
RT	9.60%	11.90%	42.90%	4.76%	-19.05%	38.10%	-23.81%	40.48%	-14.29%	-2.38%	-23.81%
Misses	7.10%	44.70%	32.90%	27.06%	-2.35%	17.65%	-42.35%	23.53%	21.18%	15.29%	-60.00%

*The minus means the color combination caused the RT prolong and the misses increase.

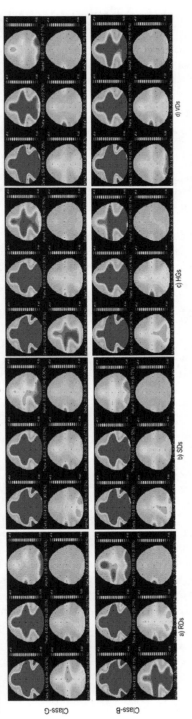

Fig.3 Electrode maps for EEG power spectra depicting the result of t-tests of the difference among class-G and class-B. Individual maps are shown for each of 6 frequency bands and were scaled from −5 to +5. Values above 4.3 (two darkest orange) and below -2.3 (two darkest blue) are statistically significant at $p<0.05$.

3 EVENT-RELATED-POTENTIAL (ERP) EXPERIMENT

In order to examine differences among different dial shapes, the classical P300 experimental paradigm was adapted to study brain function change with dial shape changes, as well as color changes.

3.1 Method

3.1.1 Subjects and Materials

Fifteen subjects were selected randomly from study one participants. They were required to sit in an armchair in an electrically shielded, sound-attenuated room.

Two kinds of combinations were selected as target stimuli respectively, one kind with shorter RT and higher accuracy named class-G, the other with longer RT and lower accuracy named Class-B.

3.1.2 Procedure

Eight trials were involved in this task. Class-G and class-B were presented as the target stimuli and non-target stimuli, respectively. In each trial target stimuli and non-target stimuli appeared randomly and the ratio between the appearance times of target stimuli and non-target stimuli was 1:4 (Zhao 2010). Subjects were required to press specified keys when the target stimuli appeared. The specified keys changed with the trial change.

Neuroscan Scan 4.5 system was adapted to record brain activity. Fourteen electrodes were placed on Fz, FCz, T3, C3, Cz, C4, T4, CPz, P3, Pz, P4, O1, Oz and O2 based on the International 10-20 System. Two electrodes, A1 and A2, linked mastoids, and A2 was the reference. Four electrodes were placed at the corners above and below the left eye, respectively, to detect ocular artifacts.

3.2 Results and Discussion

3.2.1 P300 amplitude

Figure 4 shows the P300 amplitude distribution at frontal (Fz), central (Cz), parietal (Pz) and occipital (Oz) sites of the brain. Obviously, there is a similar trend on P300 amplitude distribution among different shapes. Parietal sites mainly charge the attention change of the people. During reading the round dials, the reading mode was not simply horizontal or vertical. Along with reading, the reading mode always varies, which leads to the amplitude of parietal site increasing. Compared with RD, the reading mode of other shapes (SD, HD and VD) is much simpler; therefore, the difference among Fz, Cz, Pz and Oz is not significant. For class-B, the specified keys changes leads to a change of amplitude of central site, which is in charge of the motion.

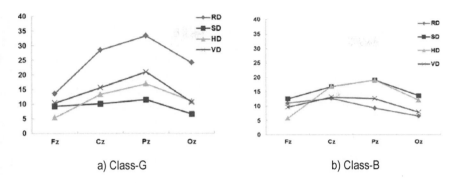

Figure 4 P300 amplitude distribution under different shapes. Fz, Cz,Pz and Oz stands for frontal, central, parietal and occipital sites of the brain respectively.

3.2.2 ERP power spectra

Six Frequency power bands were examined and defined as follows: delta (0.5~4 Hz), theta (4~8 Hz), alpha1(8~11Hz), alpha2 (11~14Hz), beta1(14~25Hz) and beta2(25~35 Hz).

Previous study has found that alpha activity reflects attentional demands and beta activity reflects emotional and cognitive processes (Ray & Cole, 1985). Alpha band power will increase during the state when people are relaxed, focused, and meditative (Cahn & Polich, 2006), while a decrease of alpha accompanies increased cognitive demand (Davis et al 2011).

In the study, warning area positions of RD and VD were all located at the upper part of the screen (see Figure 1). Alpha band power generated by RD and VD was close. Alpha band power generated by HD was larger than that of others, while that generated by SD was the least (see Figure 4). Because a horizontal reading mode is better than a vertical mode, reading HD will predictably make people feel relaxed, meanwhile the similar cognitive pattern of RD and VD leads to their alpha band power being close. Difference of the shape influence on indicators mainly reflects the warning area location and reading mode of the dials.

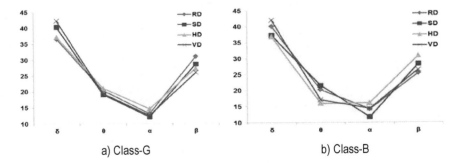

Figure 5 Power spectra changes under different shapes.

Through t-tests on power spectra of class-G and class-B, beta band power was significantly different ($p<0.05$). Figure 5 shows frontal and parietal alpha activities significantly increasing under class-B. Because parietal alpha activity can be used as an indicator of overall brain load (Holm et al 2009), reading the dials in class-B may cause visual fatigue and thus dials in class-B may need more attention to read.

The study found that an increase in brain load was associated with prolonged RT, meanwhile, prolonged RT for decision making may increase the subjects' nervousness.

4 CONCLUSIONS

In this study, through a basic psychological experiment and an ERP experiment, brain activity state on instrument cognitions was analyzed. The basic influence of instrument display characteristics related to color and shape were evaluated. Additionally EEG analysis helped to explore dynamic indexes and layout indexes et al. of the instrument design.

ACKNOWLEDGMENTS

Supported by National Natural Scientific Foundation of China (NSFC, number: NSF61075069). The authors would like to acknowledge all students in the group.

REFERENCES

Anderson ND, Habak C, Wilkinson F, Wilson HR. 2007. Evaluating shape after-effects with radial frequency patterns. *Vision Research.* 47,298-308

Bell J, Gheorghiu E, Kingdom FAA. 2009. Orientation tuning of curvature adaptation reveals both curvature-polarity-selective and non-selective mechanisms. *Journal of Vision.* 9,1-11

Bell J, Kingdom FAA. 2009. Global contour shapes are coded differently from their local components. *Vision Research.* 49,1702-1710

Cahn BR, Polich J. 2006. Meditation states and traits: EEG, ERP, and neuroimaging studies. *Psychological Bulletin.* 132,180-211

Davis CE, Hauf JD, Wu DQ, Everhart DE. 2011. Brain function with complex decision making using electroencephalography. *International Journal of Psychophysiology.* 79,175-183

Du ZF, Xu J, Zheng CX. 2009. The Study of Relationship between Cognitive Ability and Time-frequency Characteristics of ERP during Color Identification. *Chinese Journal of Biomedical Engineering.* 28,658-661

Garcia ML, Caldera CI. 1996. The effect of color and typeface on the readability of on-line text. *Computers and Industrial Engineering.* 31,519-524

Gheorghiu E, Kingdom FAA, Witney E. 2010. Size and shape after-effects: Same or different mechanism? *Vision Research.* 50,2127-2136

Holm A, Lukander K, Korpela J, Sallinen M. 2009. Estimating Brain Load from the EEG. *The Scientific World Journal.* 9,639-651

Hou YH, Zhang L, Miao DM. 2008. Physiological Influence in Vision Perception Tests with Different Color Background. *Chinese Journal of Clinical Psychology.* 16,506-508

Jiao J, Xu QL, Helander MG. 2007. Analytical modeling and evaluation of customer Citarasa in vehicle design *IEEE International Conference on Industrial Engineering and Engineering Management*, pp. 1277-81. Singapore: International Conference on Industrial Engineering and Engineering Management IEEM

Kang WY, Yuan XG, Liu ZQ. 2008. Optimization design of vision display interface in plane cockpit based on mental workload. *Journal of Beijing University of Aeronautics and Astronautics.* 34,782-785

Ma ZW, Gao SK. 2008. P300-based brain-computer interface: Effect of stimulus intensity on performance. *Journal of Tsinghua University(Science and Technology).* 48,417-420

Senol MB, Dagdeviren M, Kurt M, Cilingir C. 2009. Evaluation of Cockpit Design by Using Quantitative and Qualitative Tools. *IEEE International Conference on Industrial Engineering and Engineering Management (IEEM 2009)*, pp. 847-51. Hong Kong, PEOPLES R CHINA: International Conference on Industrial Engineering and Engineering Management IEEM

Shieh KK, Lin CC. 2000. Effects of screen type, ambient illumination, and color combination on VDT visual performance and subjective preference. *International Journal of Industrial Ergonomics.* 26,527-536

Ray WJ, Cole HW. 1985. EEG alpha activity reflects attentional demands, and beta activity reflects emotional and cognitive processes. *Science.* 228(4700), 750-752

Tilley AR, Associations HD. 2001. *The Measure of Man and Woman: Human Factors in Design*, Tianjin University Press, Tianjin, PEOPLES R CHINA

Wanyan XR, Zhuang DM, Wei HY. 2009. Pilot Attention Allocation Model In Complicated Human-machine Interface. *2nd International Conference on Biomedical Engineering and Informatics*, Tianjin, PEOPLES R CHINA, pp. 878-82.

Wei JH, Yan KL. 2008. *Basis for Cgnitive Neuroscience*, People's Education Press, Beijing, PEOPLE R CHINA

Yuan ZF, et al. 2007. *Experiment design and analysis*, China Agriculture Press, Beijing, PEOPLE R CHINA

Zeng QM, Zhuang DM. 2006. Target identification of different task weight under multi-interface and multi-task. *Journal of Beijing University of Aeronautics and Astronautic.* 32,499-502

Zeng QM, Zhuang DM, Ma YX. 2007. Design of target code in human-machine interface. *Journal of Beijing University of Aeronautics and Astronautics.* 33,631-634

Zhang L, Zhuang DM. 2009. Colormatching of aircraft interface design. *Journal of Beijing University of Aeronautics and Astronautic.* 35,1001-1004

Zhang L, Zhuang DM, Wanyan XR. 2009a. The Color Coding of Information Based on Different Mental Workload and Task Type. *ACTA ARMAMENTARII OF CHINA*.30,1522-1526

Zhang L, Zhuang DM, Yan YX. 2009b. Encoding of Aircraft Cockpit Display Interface. *Journal of Nanjing University of Aeronautics &Astronautics.* 41, 466

Zhao L. 2010. *ERP Experiment Course*. Nanjing, People's Republic of China: Southeast University Press, 108pp

Zhuang DM, Wang R. 2003. Research of target identification based on cognitive characteristic. *Journal of Beijing University of Aeronautics and Astronautics.* 29.1051-1054

Index of Authors

Abbott, R., 3
Abdulrani, M., 99
Abrahamsson, L., 135
Ajovalasit, M., 81
Akasaka, T., 177
Andreoni, G., 300

Basahel, A., 81
Bedny, G., 109, 116, 167
Bedny, I., 167
Berka, C., 269
Bone, A., 147
Breton, R., 189

Caldwell, B., 199
Carroll, M., 229
Causse, M., 157
Chafac, M., 207
Chan, A., 207
Chen, J.-J., 249
Cheng, S.-Y., 249
Cisler, D., 239
Coyne, K., 147

Dehais, F., 157

Farina, R., 43
Forester, J., 147
Forsythe, C., 3
Fukuda, T., 177

Galloway, T., 269
Gartenberg, D., 239
Gavish, N., 218
Glickman, M., 3
Gördes, D., 71
Gruppioni, E., 300
Guidi, S., 43
Guznov, S., 91

Hagiwara, H., 259
Hale, K., 229

Heemann, A., 61
Hong, J., 308

Iwasaki, H., 259

Jiang, Y., 308
Johansson, B., 135
Johansson, J., 135

Kaber, D., 71
Karwowski, W., 33, 109, 116, 125, 167
Kemp, E., 290
Keshvari, B., 99
Kokini, C., 229

Letowski, T., 53
Li, X., 308
Liao, H., 147
Lo, C.-C., 249

Marble, J., 91
McBride, M., 53
McDonald, S., 13
McGarry, R., 239
Motta, F., 300

Okada, Y., 177
Okimoto, M., 61
Oliveira, S., 61
Onal, E., 13

Parasuraman, R., 239
Parlangeli, O., 43
Perego, P., 300
Pfannenstiel, D., 239
Phan, J., 157

Ramirez, R., 290
Raphael, G., 269
Reinerman-Jones, L., 91
Rousseau, R., 189

Sanda, M.-A., 135
Satterfield, K., 290
Savoy, A., 13
Schenkman, B., 23
Ségonzac, T., 157
Shaw, T., 239, 290
Stanney, K., 229
Steffner, D., 23
Stevens, R., 269
Stoll, N., 71
Stoll, R., 71
Stripling, R., 281
Swangnetr, M., 71
Takeyama, K., 177
Tan, V., 269
Thurow, K., 71

Tremblay, S., 189
Trumbo, D., 3

Vachon, F., 189
Verni, G., 300
von Brevern, H., 33, 125
Voskoboynikov, F., 109

Wang, L., 199
Wang, W., 308
Weatherless, R., 53

Young, M., 81

Zhu, B., 71